Sustainable Business Development

Managing Change in an International Context

This open access book series focuses on the key success factors for sustainable business transformation, including the efficient use of resources, the systematic consideration of relevant aspects and technologies of digitalization, and the involvement of all relevant stakeholders. Furthermore, it identifies risks and opportunities for businesses arising from the climate crisis and digitalization and explores solutions to prevent or minimize negative effects. The series takes a multi-faceted, multi-disciplinary approach by integrating economic, sustainable, and technological perspectives. It welcomes monographs and contributed volumes from senior and promising young scholars.

Arie Hans Verkuil • Uta Milow •
Andreas Hinz • Mahmoud Al-Kilani
Editors

Core Values and Decision-Making for Sustainable Business

An International Perspective

Springer

Editors
Arie Hans Verkuil
FHNW School of Business
University of Applied Sciences and Arts
Northwestern Switzerland
Windisch, Switzerland

Uta Milow
FHNW School of Business
University of Applied Sciences and Arts
Northwestern Switzerland
Windisch, Switzerland

Andreas Hinz
FHNW School of Business
University of Applied Sciences and Arts
Northwestern Switzerland
Windisch, Switzerland

Mahmoud Al-Kilani
FHNW School of Business
University of Applied Sciences and Arts
Northwestern Switzerland
Windisch, Switzerland

ISSN 3004-9644 ISSN 3004-9652 (electronic)
Sustainable Business Development
ISBN 978-3-031-78363-0 ISBN 978-3-031-78361-6 (eBook)
https://doi.org/10.1007/978-3-031-78361-6

This work was supported by FHNW Switzerland.

This Springer imprint is published by the registered company Springer Nature Switzerland AG
The registered company address is: Gewerbestrasse 11, 6330 Cham, Switzerland

Contents

Chapter 1
Introduction: Transforming Global Challenges into Opportunities for Sustainable Entrepreneurship

Uta Milow and Andreas Hinz

Abstract This chapter introduces the thematic setting of the second volume of this book series "Sustainable Business Development" featuring 11 book chapters providing practical insights. With a rapidly growing concern for sustainability around the world, organizations face significant challenges they need to address. These challenges often relate to the concept of value at the core of organizations' business models. Moreover, stakeholder groups often are the root cause of these challenges as they can exert significant pressures on organizations to become more sustainable. Furthermore, economic and political uncertainties can be cause for concern. However, these challenges also represent opportunities, if organizations address them, align their business models and respond to expectations from stakeholders appropriately. In this context, sustainable entrepreneurship can be useful with particular emphasis on suitable solutions, target customers, market environment, supply chain processes and financial logic.

Keywords Sustainable entrepreneurship · Value · Stakeholder pressures · Sustainable business models

Today, companies operating on the international market are exposed to a variety of challenges. With the help of sustainable entrepreneurship, these challenges can be addressed and transformed into opportunities to create value and support sustainable development. In this second volume of the book series "Sustainable Business Development - Managing Change in an International Context", a wide range of national and international projects and business models is presented to provide insights into diverse contexts around the world. Typical challenges that sustainable entrepreneurship helps to address relate to the concept of value, sustainability stakeholder pressures and the political setting.

U. Milow (✉) · A. Hinz
School of Business, University of Applied Sciences and Arts Northwestern Switzerland (FHNW), Windisch, Switzerland
e-mail: uta.milow@fhnw.ch; andreas.hinz@fhnw.ch

A. H. Verkuil et al. (eds.), *Core Values and Decision-Making for Sustainable Business*, Sustainable Business Development,
https://doi.org/10.1007/978-3-031-78361-6_1

Concept of Value

An important consideration is how organizations can address sustainability challenges and create value at the same time. Sustainable business models can help organizations to holistically combine not only social and environmental activities but also economic and business-related considerations in line with the triple bottom line (Bocken et al., 2014; Elkington, 1998; Morioka et al., 2017). Sustainable business models are not limited to "businesses" and can be equally relevant for any type of organization that pursues sustainability activities, whether private or public sector and for-profit or not-for profit. Sustainable business models can support organizations to develop and implement sustainable strategies across their value chain operations and processes (Morioka et al., 2017).

Organizations have three key elements of sustainable business models at their disposal which they can work with (Morioka et al., 2017):

- Value proposition (e.g., product and service offering)
- Value creation and delivery system (e.g., capabilities, resources and networks)
- Value capture (e.g., financial and non-financial, reputation)

A key question to ask is how value in the context of sustainable business models is defined.

Purely financial value can simply be defined based on typical financial performance measures such as return on investment, profit margins, earnings per share and many others. However, when organizations seek to go beyond an exclusive economic scope to create social and environmental value, many specific measures for a given context need to be defined. Even when the three pillars of sustainability (Elkington, 1998) are considered, individual perceptions of sustainable value differ and have also evolved over time (Morioka et al., 2017). It can be observed that the main difference does not lie in how many of the three pillars of sustainability are taken into account, but rather in the prioritization among them. When the notion of value is dominated by economic goals, sustainable development is seen as means to achieve monetary goals or social and environmental performance measures are expressed in financial terms (Ang & Van Passel, 2010; Hart et al., 2003; Morioka et al., 2017). For a more balanced integration of social, environmental and economic aspects, the focus shifts towards how well organizations satisfy different needs and expectations of respective stakeholders (Morioka et al., 2017).

Sustainability Stakeholder Pressures

With sustainability considerations playing an increasingly important role, organizations encounter significant pressure to become more sustainable. These pressures originate from various internal and external stakeholder groups and organizations face the challenge of addressing their specific needs (Bello-Pintado et al., 2023; Journeault et al., 2021). In dealing with these challenges, organizations seek to create value, and these activities may affect or be affected by relationships organizations have with their respective stakeholders (Góes et al., 2023). Stakeholder theory suggests that these interactions have a direct impact on how organizations perform (Góes et al., 2023; Phillips et al., 2019). To engage appropriately with stakeholders, organizations therefore have an intrinsic interest to understand exactly who their stakeholder groups are and which specific needs they have (Góes et al., 2023; Phillips et al., 2019). In this regard, a central tenet of stakeholder theory is that value is created for each stakeholder group without compromising another group (Freeman, 2010; Góes et al., 2023).

Depending on the field of activity and industry sector, the public profile and general perception of organizations as well as the power that these stakeholder groups exert may vary significantly (Bello-Pintado et al., 2023). For instance, this is determined by factors such as the type and quality of relationship of an organization with a given stakeholder group or the urgency of the matter (Bello-Pintado et al., 2023; Freeman, 1994; Mitchell et al., 1997). Following the classification of Bello-Pintado et al. (2023), external stakeholder groups include customers, government and activist groups while internal stakeholder groups include owners-shareholders, employees and managers.

A typical example of stakeholder expectations is that organizations become more sustainable by taking into account the United Nations Sustainable Development Goals (SDGs) in their activities (Blasi et al., 2021; Tyler et al., 2024). While such stakeholder pressures are often perceived as threats, they also represent significant opportunities (Morioka et al., 2017; Tyler et al., 2024). Along these lines, organizations can benefit from implementing sustainability practices in multiple ways. While they can satisfy stakeholder needs and therefore improve their public image and differentiation, they can also capture resource efficiency gains in operational processes across their value chain (Muñoz & Cohen, 2018; Tyler et al., 2024). Adopting more sustainable practices can therefore reduce negative impact while at the same time increasing financial and reputational capital (Blasi et al., 2021; Tyler et al., 2024).

Adapting the business model towards sustainability considerations is widely seen as an opportunity (Morioka et al., 2017; Tyler et al., 2024). Nevertheless, organizations often face difficulties when streamlining operations in favor of sustainability. They may encounter trade-offs when attempting to become more sustainable and these have an impact on actual sustainability practices (Morioka et al., 2017). For instance, it is quite common that sustainability initiatives are perceived as additional cost drivers, especially when benefits are expected to take longer to materialize (Tyler et al., 2024).

Political Setting

The USA has been promoting international free trade since the end of the Second World War, which was developed on the three pillars of the World Bank, the Monetary Fund and the World Trade Organization. This global economic policy approach was combined with a connection to a pluralistic political value system (Felbermayr et al., 2017). Since then, however, various global tensions and diverging national interests have resulted in protectionist tendencies between various countries and geographic regions across the world. This comes at a significant cost because cross-border trade in goods and services as well as foreign direct investment have contributed significantly to economic growth and prosperity worldwide for decades. International trade and investments have supported the global economic integration of developing countries and reduced poverty. Global labor mobility and division of labor have caused considerable productivity gains among participating companies. It has stimulated innovation and led to greater product diversity (Hilpert & Rudloff (2024); Hilpert, 2022).

Addressing These Challenges

Against this backdrop, international entrepreneurship strategies can help to capture the potential that comes with these challenges and to increase the chances of genuine and sustainable value creation. Sustainable entrepreneurship in particular can act as a catalyst for the development and implementation of effective solutions to address the abovementioned challenges on a global scale. This potential is by no means limited to the entrepreneurial ecosystem focusing on start-ups. Because entrepreneurial tools and approaches can also be applied in the context of more mature and larger companies from Small and Medium-sized Enterprises (SMEs) to Mutli-national Enterprises (MNEs), which is often referred to as intrapreneurship and corporate entrepreneurship, their potential impact can be multiplied considerably. And it goes even further, as these tools and approaches can be useful for any type of organization, from public/governmental institutions to non-governmental organizations and others.

For sustainable entrepreneurship to lead to suitable and competitive solutions that can create real impact, several considerations should be taken into account:

- General market environment and competition
- Target customers and market entry
- Operational processes and supply chains
- Financial logic

General market environment and competition: Various conditions need to be considered when entering new markets. The first step is to conduct in-depth market research on the target market. This requires an understanding of market characteristics and knowledge of the potential demand for the services and products on offer. A new competitive landscape is also always associated with the cultural, political and legal characteristics of the target market.

Target customers and market entry: Regarding market positioning and reaching target customers, entrepreneurship will also have to pay attention to market specifics such as digital presence. How are online marketing, social media and e-commerce platforms used in the target market to reach customers? Does the company have enough qualified employees in terms of international business practices, language and intercultural competence? These are just exemplary aspects that need to be carefully integrated into the planning process in order to successfully enter international markets and pursue a sustainable business in the long term (Weerawardena & O'Cass, 2009).

Operational processes and supply chains: Another prerequisite for sustainable entrepreneurship is to establish adequate production methods and supply chains. This may include the implementation of fair working conditions, sourcing from organic or fair-trade suppliers and the social integration into the local community. It is therefore essential to establish partnerships with local organizations that have market knowledge and networks. These measures are also essential for establishing effective distribution channels and logistics partnerships in the target market (Yumei et al., 2021).

Financial logic: The issue of sufficient financial resources as part of entrepreneurial risk management is often underestimated. It is essential to allocate sufficient financial resources for international market entry and expansion, logistics and operations as well as exchange rate risks or currency restrictions (Bishev & Boskov, 2016).

To provide real-world practical insights into how this is done, 11 contributions have been selected as chapters in this book. Part one deals with more fundamental topics like core values, while part two gives insights in sustainability challenges and improvements from various countries.

In his article "Sustainability in SMEs: what core values are really important?", Stéphano Vacher examines the social context of an SME-based corporate policy. He describes the commitment to corporate social responsibility and the relationship between managers and their stakeholders in relation to CSR engagement. According to his research, environmental, social and societal issues appear to be addressed differently depending on the strategic and ethical values of the manager. These values can promote CSR engagement and ensure the company's growth while contributing to global well-being. The author collected the data from two different surveys.

The authors Giorgio Gatica Singgih and Melia Famiola Hariadi from Bandung present an Indonesian study. It is a challenge for companies in Indonesia to implement sustainability standards. This is especially true for SMEs. The authors describe their research findings on how environmental, social and governance parameters can be implemented in SMEs.

In the next chapter, Dina Dellyana and Leo Aldianto examine the importance of sustainability in modern marketing strategies. These are driven by increasing consumer awareness and regulatory requirements. The study examines how companies are integrating environmentally friendly practices to improve their brand reputation and ensure their long-term profitability. In doing so, they find that companies practicing green marketing can improve their market position. The study also highlights the importance of environmental responsibility, social responsibility and transparency in marketing.

Irayanti Adriant, Togar Mangihut Simatupang and Yuanita Handayati describe the advantages of a sustainable business approach for companies in an Indonesian study on agricultural supply chains. The research emphasizes that sharing costs can help reduce the financial strain on farmers, allowing them to invest in sustainable practices and boost their profits. It also examines how incentives from retailers can encourage farmers to adopt improved farming methods, resulting in higher-quality crops and greater profits for everyone involved. Farmers are enabled to produce both more cost effective and more sustainable.

The authors Minu Mehta and Boishampayan Chatterjee analyze the role of micro-enterprises in India with regard to the introduction of sustainable business practices. According to the authors, environmental, social and governance (ESG) parameters are playing an increasingly important role. They present a new concept that can be of decisive importance for the introduction of ESG. Their multifactorial model is to become the basis for implementing and measuring ESG criteria in Indian micro-enterprises.

In their article, Wawan Dhewanto and Rozan Hanifan examine the factors that influence sustainability initiatives by micro-enterprises. Their study focuses on the Indonesian market and therefore offers an ideal comparison to the study on micro-enterprises in India by the authors Minu Mehta and Boishampayan Chatterjee. This article examines the role of stakeholder participation in promoting sustainable business practices among MSMEs in Indonesia. Both the results of a quantitative survey and the results of stakeholder interviews are used. This research contributes to highlighting the importance of internal versus external factors in promoting the sustainability of micro-enterprises.

In the article for the Swiss market context, Karsten Eichler, Dario Meyer and Rolf Meyer examine the question of why, in fact, only a small number of companies are still innovating business models. Entrepreneurial success plays a key role in sustainable economic growth, benefiting society at large. Over time, it helps maintain economic stability, generates employment, and encourages innovation, all of which contribute to a flourishing economy. Successful businesses are more inclined to invest in sustainable practices, support community initiatives, and promote social progress. This study found that the innovation rate of companies in Switzerland is generally declining, especially among small and medium-sized enterprises (SMEs). The authors describe the obstacles to business model innovation among small and medium-sized enterprises in Switzerland. 4000 companies were surveyed in a written questionnaire. The results show that medium-sized companies and those in high-tech sectors are more inclined to pursue business model innovation. In contrast, smaller companies and companies in less technical sectors face greater challenges and obstacles, which are highlighted in the article.

The authors Ivan Köhle and Beat Birkenmeier have taken a different approach and go back to the roots of sustainable international success—education. In order to be able to realize sustainable entrepreneurship on international markets at all, universities are required to adapt their curricula to the new challenges. The authors present a study on the extent to which the business administration faculties of Swiss universities already meet this requirement in their bachelor's degree programs and where there are still considerable gaps. The second part of the paper presents the integrated didactic approach of an internationalization cube implemented at the University of Applied Sciences Northwestern Switzerland (FHNW) and its effects.

In the following article, Agnès Walser-Luchesi, Anne-Catherine Furst and Landisoa Rabeson analyze the consumer behavior of young people from generations Y and Z with regard to the purchase of luxury products and the commitment of luxury companies to the sustainability of their products. They show that attitudes towards second-hand luxury goods, subjective norms and perceived behavioral control have a significant influence on purchase intentions. In addition, the quality of pre-owned luxury goods is considered important, but the aspect of conscious sustainable consumption remains of secondary importance for the target group.

Daniel Borner's and Barbara Eisenbart's study on a circular economy project in the U.S. concludes our compilation of projects and studies on successful sustainable business strategies. The authors describe the interplay of so-called hard and soft methods in the circular economy. While the hard methods focus digitalization and new technologies with investments and operating costs, the soft methods include collaboration and knowledge sharing. These methods do not require any initial investment but rely on learning from partners to improve their processes and collect data to evaluate internal structures.

In the final chapter, the editors summarize today's challenges for SME with cost pressures, sustainability pressures of the stakeholders and uncertainties in the political and economic ecosystem.

References

Ang, F., & Van Passel, S. (2010). The sustainable value approach: A clarifying and constructive comment. *Ecological Economics, 69*(12), 2303–2306. https://www.sciencedirect.com/science/article/pii/S0921800910002223

Bello-Pintado, A., Machuca, J. A., & Danese, P. (2023). Stakeholder pressures and sustainability practices in manufacturing: Consideration of the economic development context. *Business Strategy and the Environment, 32*(7), 4084–4102.

Bishev, G., & Boskov, T. (2016). Financial impact strategy on SMEs in the business world. *International Journal of Current Research, 8*(9), 39226–39229.

Blasi, S., Crisafulli, B., & Sedita, S. R. (2021). Selling circularity: Understanding the relationship between circularity promotion and the performance of manufacturing SMEs in Italy. *Journal of Cleaner Production, 303*, 127035.

Bocken, N. M., Short, S. W., Rana, P., & Evans, S. (2014). A literature and practice review to develop sustainable business model archetypes. *Journal of Cleaner Production, 65*, 42–56.

Elkington, J. (1998). Partnerships from cannibals with forks: The triple bottom line of 21st-century business. *Environmental Quality Management, 8*(1), 37–51.

Felbermayr, G., Steininger, M., & Yalcin, E. (2017). *Konsequenzen einer protektionistischen Handelspolitik der USA (No. 88)*. ifo Forschungsberichte.

Freeman, R. (1994). The politics of stakeholder theory: Some future directions. *Business Ethics Quarterly, 4*(4), 409–421.

Freeman, R. E. (2010). *Stakeholder theory: The state of the art*. Cambridge University Press.

Góes, H. A. D. A., Fatima, G., Santos Jhunior, R. D. O., & Boaventura, J. M. G. (2023). Managing for stakeholders towards corporate environmental sustainability. *Corporate Social Responsibility and Environmental Management, 30*(4), 1561–1572.

Hart, S. L., Milstein, M. B., & Caggiano, J. (2003). Creating sustainable value. *Academy of Management Perspectives, 17*(2), 56–69. https://doi.org/10.5465/AME.2003.10025194

Hilpert, H. G. (2022). *Zeitenwende in der EU-Handelspolitik: Chancen der Diversifizierung im Indo-Pazifik.*

Hilpert, H. G., & Rudloff, B. (2024). EU-Handelspolitik: Die neue Nachhaltigkeitsfalle für handelspolitische Partnerschaften. *Ifo Schnelldienst, 77*(1), 3–7.

Journeault, M., Perron, A., & Vallières, L. (2021). The collaborative roles of stakeholders in supporting the adoption of sustainability in SMEs. *Journal of Environmental Management, 287*, 112349.

Mitchell, R. K., Agle, B. R., & Wood, D. J. (1997). Toward a theory of stakeholder identification and salience: Defining the principle of who and what really counts. *Academy of Management Review, 22*(4), 853–886.

Morioka, S. N., Bolis, I., Evans, S., & Carvalho, M. M. (2017). Transforming sustainability challenges into competitive advantage: Multiple case studies kaleidoscope converging into sustainable business models. *Journal of Cleaner Production, 167*, 723–738.

Muñoz, P., & Cohen, B. (2018). Entrepreneurial narratives in sustainable venturing: Beyond people, profit, and planet. *Journal of Small Business Management, 56*, 154–176.

Phillips, R. A., Barney, J. B., Freeman, R. E., & Harrison, J. S. (2019). Stakeholder chapter. *The Cambridge handbook of stakeholder theory*, 3.

Tyler, B. B., Lahneman, B., Cerrato, D., Cruz, A. D., Beukel, K., Spielmann, N., & Minciullo, M. (2024). Environmental practice adoption in SMEs: The effects of firm proactive orientation and regulatory pressure. *Journal of Small Business Management, 62*(5), 2211–2246.

Weerawardena, J., & O'Cass, A. (2009). Examining the role of international entrepreneurship, innovation and international market performance in SME internationalization. *European Journal of Marketing, 43*(11/12), 1325–1348.

Yumei, H., Iqbal, W., Nurunnabi, M., Abbas, M., Jingde, W., & Chaudhry, I. S. (2021). Nexus between corporate social responsibility and firm's perceived performance: Evidence from SME sector of developing economies. *Environmental Science and Pollution Research, 28*, 2132–2145.

Uta Milow, Prof. Dr., is a lecturer and researcher of economics at FHNW with a research focus on sustainable and international entrepreneurship. She attended the University of Mannheim, Germany, and the University of Oregon, USA. After finishing a master's degree in economics at the University of Mannheim, she continued her studies with a PhD-program in Economics at Eberhard-Karls-University Tübingen, Germany. She participated in an interdisciplinary postgraduate research program on European integration (political science, law, and economics). Since 2000 she has been a lecturer at the School of Business FHNW. She is head of the Master of Business Administration (MBA) programs at the School of Business, FHNW, and is responsible for all executive education programs of the Institute of Management, FHNW.

Andreas Hinz, Prof. Dr., is lecturer and researcher at the Institute of Management, School of Business, University of Applied Sciences and Arts Northwestern Switzerland. He focuses on (Sustainable) Entrepreneurship and Innovation as well as Management. He is co-founder of Sustinova, a sustainable venture, and works as mentor at SINGA Switzerland, a startup incubator for refugees. Andreas holds a PhD in Sustainability Innovation and has done postgraduate studies in Entrepreneurship. He has work experience in Asset Management, Sustainability, and Philanthropy.

Part I
Core Values of Sustainable Business Decision-Making

Chapter 2
Sustainability in SMEs: What Core Values Are Really Important?

Stéphano Vacher

Abstract There is a genuine commitment to Corporate Social Responsibility (CSR) among companies of all sizes. However, for small and medium-sized enterprises (SMEs), there is a distinct relationship between managers and their stakeholders in terms of CSR commitment. Environmental, social, and societal issues are approached differently depending on the strategic and ethical values held by the manager. These values can enhance CSR commitment, ensuring the firm's growth while contributing to global well-being. This paper focuses on this relationship through two studies: (1) a descriptive survey conducted in 2019 on a sample of French companies, examining their responses to CSR commitments, and (2) a qualitative survey of 10 French SME managers, exploring the criteria influencing their CSR engagement. The findings reveal that CSR actions are strongly influenced by the strategic goals and values upheld by the firm. SMEs have their own specific management style. In fact, the way they operate responds to a number of mechanisms that are complex to identify. Relationships with stakeholders in the firm's ecosystem, combined with the values espoused by the firm's managers, can be decisive factors in determining action in favour of CSR.

Keywords SME · CSR · Stakeholder theory · Core values · Environmental · Social

Introduction

Corporate Social Responsibility (also called CSR) has emerged as a fundamental pillar of contemporary business strategy, reflecting a paradigm shift towards sustainable and ethical practices (Bowen, 2013). In today's rapidly evolving global landscape, businesses are increasingly expected to go beyond mere profit generation

S. Vacher (✉)
EM Strasbourg Business School, Université de Strasbourg, Strasbourg, France
e-mail: stephano.vacher@em-strasbourg.eu

A. H. Verkuil et al. (eds.), *Core Values and Decision-Making for Sustainable Business*, Sustainable Business Development,
https://doi.org/10.1007/978-3-031-78361-6_2

and actively contribute to societal well-being and environmental preservation. This necessitates a deeper exploration into the multifaceted dimensions of CSR engagement, encompassing not only its strategic implications but also its ethical and moral imperatives.

The integration of CSR into organizational frameworks goes beyond a superficial commitment; it requires a profound alignment with the core values and mission of the company. When CSR initiatives are deeply rooted in the organizational DNA, they cease to be peripheral activities and instead become intrinsic to the firm's identity and purpose. This alignment not only enhances the company's reputation and stakeholder trust but also fosters long-term sustainability by ensuring that social and environmental considerations are woven into every aspect of its operations.

Moreover, the cultural context within which CSR operates plays a pivotal role in shaping its implementation and effectiveness. Organizational culture, defined by shared beliefs, values, and norms, significantly influences managerial behavior and decision-making regarding CSR initiatives (Davis, 1960; Swanson, 2008). Understanding how cultural values impact CSR commitment is essential for devising strategies that resonate with employees and stakeholders alike, driving meaningful change within and beyond organizational boundaries.

In addition to cultural considerations, the legitimacy of CSR efforts is crucial for their success and impact. Legitimacy, derived from societal norms and expectations, serves as a foundation upon which firms build trust and credibility with their stakeholders. By aligning CSR initiatives with societal values and expectations, companies not only enhance their reputation but also mitigate risks and navigate regulatory complexities more effectively, ultimately contributing to long-term business resilience and sustainability.

Furthermore, the integration of CSR into core business activities requires a strategic approach that goes beyond philanthropy or token gestures (Courrent, 2012; Marchesnay, 1993). Sustainable business practices, inclusive supply chains, and innovative solutions to societal challenges are some examples of how companies can leverage their operations to drive positive change while simultaneously creating value for stakeholders. By embedding CSR into the fabric of their business models, companies can unlock new opportunities for growth, differentiation, and competitive advantage in an increasingly socially conscious marketplace.

In essence, a holistic understanding of CSR entails examining its strategic, ethical, and societal dimensions in tandem. By delving deeper into the interplay between organizational mission, culture, legitimacy, and business activities, researchers and practitioners can develop more nuanced insights and strategies for fostering meaningful CSR engagement. This requires not only a commitment to corporate citizenship but also a recognition of the profound impact that businesses can have on society and the environment when they align profit motives with social and environmental responsibility.

The stance adopted by managers actively engaged in Corporate Social Responsibility (CSR) holds significant implications for safeguarding a firm's operations and nurturing business relationships. The managerial conceptualization of sustainability traces its origins to Bowen (2013). CSR has progressively solidified its presence in business practices, facilitated by regulatory frameworks fostering societal engagement in communal issues. Carroll (2016) delineates the manager's responsibilities within the CSR pyramid, encompassing economic, legal, ethical, and philanthropic dimensions. Fel (2011) underscores that companies have recognized resource management as a pivotal factor in ensuring the sustainability of their operations. Themes such as environmental sustainability, social considerations, and comprehensive production chain control are gaining heightened significance for corporations (Lu et al. 2018). Dahlsrud (2008) demonstrates a broad consensus in existing CSR definitions. The author argues that the primary challenge is not in defining Corporate Social Responsibility (CSR), but in understanding how CSR is socially constructed within particular contexts. He asserts: *"Definitions of CSR describe a phenomenon, but give no guidance on how to deal with the challenges associated with this phenomenon. Therefore, the challenge for companies is not so much to define CSR as to understand how CSR is socially constructed in a specific context and how to take this into account when developing business strategies."* (Dahlsrud, 2008, p. 6). According to Dahlsrud (2008), five dimensions hold paramount importance: environmental, social, economic, stakeholder, and voluntary.

The embracement of Corporate Social Responsibility (CSR) and sustainable practices is evidently becoming more pervasive within a growing spectrum of companies. The process of selecting and referencing suppliers within the framework of a sustainability strategy is increasingly transitioning from theoretical to operational phases in French firms. Academic literature underscores the significance of site certification, adherence to rights and working conditions, and the establishment of an environmental charter. Recent academic inquiries have aimed at elucidating the motivators and risks associated with adopting a sustainable approach. Primary stakeholders are assuming an expanding role in catalysing responsible initiatives (Yu & Ramanathan, 2015). Nonetheless, a comprehensive engagement with partners by all stakeholders remains elusive, as evidenced in the study excerpt: "*... On the supplier side, 39% say that CSR is important to their buying customers only on paper but does not translate into concrete actions ....*". Consequently, delineating the sincerity or strategic underpinnings of corporate commitment mechanisms emerges as a nuanced challenge.

Stakeholder theory (Freeman, 2010) and the models proposed in the existing literature could be used to better understand the subject. However, studies analysing the commitment of SMEs to CSR issues remain limited (Boakye et al., 2020; Chassé & Courrent, 2018; Graafland, 2018; Zhu et al., 2019).

In a systematic review of the literature on the topic of CSR from 2000 to 2016 representing 14,490 articles in 15 CSR journals, we note that most of the main authors of research on CSR have been published in the context of English-speaking territories (Lin et al., 2020). Thus, the research topics are overwhelmingly focused on the aforementioned context. It would be interesting to offer a new perspective to fill the academic gap by studying the French context.

Listed companies have a very small numerical workforce in France, whereas small and medium-sized enterprises (also called SMEs) represent the largest part of the active population on French territory. The economic and financial weight of listed companies strongly conditions the growth and development of their sector of activity, as well as the global economy as a whole. For their part, SMEs contribute to the smooth functioning of the global value chain. They also represent for the majority of companies in the world (Levratto, 2009).

There are 159,000 SMEs in France, employing 4.3 million people and generating 23% of the added value of all businesses in 2021.[1] Spence (1999) and others (Jenkins, 2004; Madsen & Ulhøi, 1996; Spence et al., 1998) notes that sometimes the CSR-oriented actions carried out by SMEs present notable specificities that must be taken into account to better understand their commitment (Jamali et al., 2009).

Our research proposal will shed light on each criterion of Courrent (2012) and Marchesnay (1993) model, focusing on values as the culture, legitimacy, know-how and mission in according with CSR involvement (Gherib et al., 2009).

Our objective, in the first place, will be to present the societal commitment of companies and managers in a discrete descriptive study from the INSEE survey (ENDD, 2016) published in 2019, concerning CSR commitment.

In a second step, we will see how SMEs managers perceive the criteria the culture, legitimacy, know-how and mission in terms of reflection and in terms of action when these are linked to CSR commitment.

The analysis is based on the data collected in our survey developed with 10 managers during the year 2024. The problematic will answer several questions of an exploratory nature:

- **What criteria emerge from the posture of CSR managers?**
- **Why do they implement these values and what actions do they take?**
- **Which CSR pillar is favoured according to the value criterion chosen?**

Our work questions the reasons that drive SMEs managers to take action as part of their CSR strategy. We have therefore chosen to mobilise both the stakeholder theory and the Marchesnay model, whose complementarity will make it possible to highlight the influence of criteria on the behaviour of managers in CSR commitment.

The first part recalls the stakeholder theory (Freeman, 2010), CSR definitions completed by the Marchesnay model (Marchesnay, 1993) revisited by Courrent (2012). The second part presents the method, the third part presents the results and the fourth part discusses the results obtained. The conclusion will serve as the last part.

Theoretical Approach

The Stakeholder Theory

- Since the dissemination of the seminal approach initiated by Freeman (2010) and the definition of the stakeholder concept, the literature mobilising this theory has expanded in its treatment of the subject of CSR. Stakeholder management is

[1] https://www.insee.fr/fr/statistiques/7678534?sommaire=7681078

fundamental to the long-term development and growth of the company. Indeed, the stakeholders thus presented can be affected by the firm's activities. There has been a succession of authors who have tried to understand this concept. The literature on stakeholder theory sometimes crosses with firm-centred strategy (Freeman, 2010; Hill & Jones, 1992; Vandchali et al., 2021), or ethics centred on the interaction between the firm and its stakeholders (Park-Poaps & Rees, 2010).
- We find it relevant to analyse existing inter-organizational relationships in the light of stakeholder theory as indicated by Freeman (2010). Lozano et al. (2015) consider it interesting to mobilise stakeholder theory to explain the relationships that develop between a firm and its stakeholders. In a broad sense, the stakeholders present in the firm's environment push the firm to act to ensure its survival and development (Freeman, 2010; Wood et al., 2021). Companies have been engaged for more than a decade in practices to improve their environmental, societal and governance practices to meet societal expectations (Yawar & Seuring, 2018). This fundamental movement affects all the actors present in the firm's value chain, which can constitute in passing a performance lever that can benefit all the actors in the ecosystem. The firm's interest is sometimes mainly motivated by cost reduction (Dyck & Silvestre, 2018). Nevertheless, stakeholder action can lead to the improvement of the production process (Sandin & Peters, 2018). Primary stakeholders are essential for the survival of companies and include suppliers, customers, employees and shareholders (Clarkson, 1995).

Corporate Social Responsibility (CSR)

- CSR represents the transposition of the sustainable development approach applied to the business world. This concept is under construction, so much so that the attention paid to it goes beyond the purely economic sphere. The concept is obviously not new (Bowen, 2013; Carroll, 2016; Carroll & Buchholtz, 2015), but its lineage is in constant evolution.
- Bowen (2013) was the first author to define what society's expectations of the businessman might be, beyond the financial intuitions cherished by Friedman (1953). Friedman (1970) justifies his position in capitalism and freedom by stating, "*... there is one and only one social responsibility of business - to use its resources and engage in activities designed to increase its profits so long as it stays within the rules of the game, which is to say, engages in open and free competition, without deception or fraud.*" Friedman believes that the manager's only responsibility is to make a profit, as indicated by the evocative title of his famous article published in the New York Times on 13 September 1970: "The Social Responsibility of Business Is to Increase Its Profits".
- Some studies show that the search for competitiveness is one of the driving forces behind commitment to a CSR approach (Bansal & Roth, 2000; Juholin, 2004; Porter & Van Der Linde, 1995a, b).

- Following Bowen, the authors emphasise societal commitment in the broadest sense: Legal (Chamberlin, 1953), legitimacy (Crossley et al., 2021), public accountability and managerial discretion (Wood, 1991), total socio-economic welfare (Frederick, 2018), responsibility to society that goes beyond legal obligations (McGuire, 1969), social responsibility begins where the law ends (Davis, 1973), beyond legal or contractual requirements, organizations have an obligation to societal actors (Jones, 1999), the triple bottom (Carroll, 2016; Elkington, 1998), or ethics (Swanson, 1995). All authors agree that the relationship with stakeholders is essential to successfully meet the challenges posed by CSR.
- As mentioned above, the multiplicity of definitions, perspectives and literature on CSR makes it a complex task to determine where, how and when the company can objectively contribute to the good of society (Aguinis & Glavas, 2012; Wang et al., 2020).
- Table 2.1 presents a chronological list of some of the definitions of CSR proposed by academic authors and business actors.

CSR-oriented Strategic Management in SMEs

- In his book, Marchesnay (1993, p. 251) proposes a method of strategic analysis based on a matrix incorporating a number of variables.

Table 2.1 Few definitions of CSR

References	Definitions of Corporate Social Responsibility (CSR)
Bowen (2013)	Corporate social responsibility refers to the obligation of business leader to make policies, take decisions and follow courses of action that meet the objectives and values that are considered desirable in our society.
Carroll (1979)	Corporate social responsibility integrates all the economic, legal, ethical and discretionary expectations that society may have of a company at a given time.
Jones (1980)	The idea that organisations, beyond legal or contractual requirements, have an obligation to societal actors.
Wartick and Cochran (1985)	Societal responsibilities are defined by society, and the company's activities are: (a) identify and analyse the changing expectations of society in relation to the responsibilities of the company; (b) determine a comprehensive approach to be responsible for the changing demands of society; (c) to implement appropriate responses to the social problems encountered.
European Commission (2011)	This is the responsibility of companies with regard to the effects they have on society. In order to fulfil this responsibility, companies must first comply with existing legislation and collective agreements between social partners. In order to fulfil their social responsibility, companies should have engaged in a process, in close cooperation with their stakeholders, to integrate social, environmental, ethical, human rights, and consumer concerns into their business operations and core strategy.

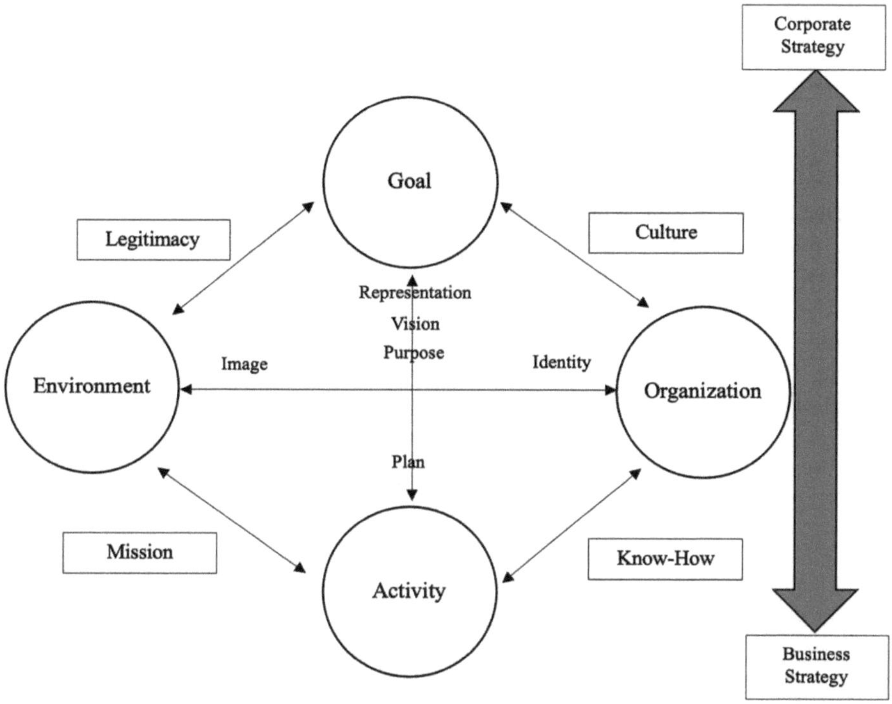

Fig. 2.1 Adapted from Courrent (2012) and Marchesnay (1993)

- This approach makes it possible to move beyond a logic that is strongly based on a highly framed rationality and instrumental approach, towards a decision structure based on a limited rationality that mobilises, in particular, the processes and organizational dynamics of the firm.
- Courrent (2012, p. 64), remobilises the strategic approach to adapt it to the issues raised by CSR (Fig. 2.1). The author indicates that it is essential to integrate a broader vision of the information that managers need to guarantee the firm's economic and financial performance (Paradas, 2007). It also leads to a global performance approach, in particular by taking into consideration the notion of risk. Lastly, the methods used must make it likely to grasp the specific issues linked to sustainable development and CSR, by organizing dialogue with stakeholders (Berger-Douce, 2006; Hammann et al., 2009; Lapointe & Gendron, 2005). This is explained in Fig. 2.1.
- Thiétart (2014, p. 200) explain that: "for a given conceptual definition, there is no empirical data corresponding exclusively to that concept. Similarly, a researcher who wishes to make the transition from the empirical to the theoretical world has conceivable elements as the manifestation of several potential concepts."
- Therefore, we propose to study the impact of this sub-criterion in the theoretical model Courrent (2012) adapted from Marchesnay (1993). We will mobilise the latter in our research work. We will thus be able to shed light on and refine the CSR manager's perceptions in their involvement in CSR.

Research Methodology

We first present the CSR commitment of French companies. Secondly, we developed a survey of 10 managers in order to highlight the factors that influence these same issues.

A descriptive survey aims to understand the relevance of a phenomenon and to describe its distribution within a given population. Forza (2002, p. 155) states that "descriptive survey research aims to understand the relevance of a certain phenomenon and to describe the distribution of the phenomenon in a population. Its primary objective is not theory development, although through the facts described, it can provide useful insights for both theory development and refinement (Malhotra & Grover, 1998; Wacker, 1998). In our paper, a quantitative descriptive study based on INSEE data is conducted to determine whether or not engagement on economic and societal issues is a reality among French companies. Then, we extend our research work by presenting the qualitative survey conducted on a population of 10 French Sme's managers.

Secondary Data

Source of Additional Data to Analyse CSR and Corporate Social Responsibility

As mentioned by Ellram and Tate (2016, p. 250), secondary data is becoming increasingly important in management science research and can be an interesting additional data source. Secondary data sets often use well-established measures that add credibility when combined with the results of another study (Ellram & Tate, 2016, p. 251). The use of secondary data allows procurement researchers in supply management (Ellram & Tate, 2016) to triangulate the results of primary data collection, such as interviews, for example. For our research, secondary data helps us to determine whether CSR involvement or commitment is a reality among many companies. A population base representing 80,445 companies was used to determine CSR commitment. The data are from a survey conducted by INSEE in May 2019.[2] Legal units with 500 or more employees are surveyed exhaustively, 2158 of a total sample of 11,009 units.[3] Participation in the survey is free for firms with fewer than 500 employees, The units surveyed are legal units with at least 20 employees. Secondary data must also have contextual validity to be used in a meaningful way (Ellram & Tate, 2016).

[2] https://www.insee.fr/fr/metadonnees/source/fichier/MethodesEtPratiqueDesEnquetesEntreprises.pdf

[3] https://www.insee.fr/fr/metadonnees/source/operation/s1306/processus-statistique

Then, from the 11,009 units we extracted raw data obtained from the DSAC and reprocessed by ourselves for the 8997 companies analysed. The INSEE study was developed from the Sirus business register and aims to establish an inventory of companies engaged in sustainable development and corporate social responsibility (CSR) by taking into account different dimensions: environmental, economic and social. The 55 questions used in the INSEE study are closed questions (yes, no, don't know/not concerned or, in some cases, multiple choice). The information collected is therefore strictly quantitative and discontinuous. The INSEE survey is considered a well-established data source. We mobilise 1 question in the study to measure CSR engagement using the definition of CSR proposed by the European Commission (2011): *"the responsibility of enterprises for their impacts on society"*, the INSEE study asks companies about their societal commitment. Thus, defined for the respondents, the concept of CSR and societal commitment to suppliers and customers is then addressed in the survey in its operational form through the following questions:

Q1. Do you consider yourself to be a manager involved in Corporate Social Responsibility (CSR) or Sustainable Development (SD)? (Answer: Yes/No).

It also addresses issues related to the commitment to CSR. This survey provides an overview of a broad area that will allow the impact of public policy development on this involvement to be assessed and guided in the future. It responds to a need for information on the part of the ministry responsible for their animation, intermediary organizations and NGOs.

The data collected was analysed using descriptive statistics. Additional data extracted from INSEE allowed us to believe that the element that apparently really counts in CSR involvement is the size characteristic of companies.

Primary Data

Method of Collecting and Analysis Primary Data

The data was collected through a quantitative and qualitative survey self-administered of a sample of 10 managers working in France including 2 women and 8 men. Table 2.2 presents a profile of the 10 respondents. We analysed the part quantitatively by means of the descriptive study to expose the distribution of the proportions of respondents with regard to the quantitative questions proposed in our study. We again use the methodology outlined above.

Analysis of Qualitative Data

We then analysed the data qualitatively, using the method of thematic content analysis, in order to be able to synthesise all the data collected, as recommended by recognised authors (Feller, 1977; Rockmann & Vough, 2023). This analysis technique

Table 2.2 Profile of respondents

Respondent's code	Respondent's position	Sector	Employees	Gender
Resp 1	CEO shareholder	Services	28	Female
Resp 2	CEO shareholder	Services	78	Male
Resp 3	CEO employee	Construction	250	Male
Resp 4	CEO shareholder	Construction	20	Male
Resp 5	Manager employee	Services	230	Male
Resp 6	CEO shareholder	Funeral company	48	Male
Resp 7	Manager employee	Metallurgy	200	Male
Resp 8	CEO shareholder	Manufacturer	14	Male
Resp 9	CEO shareholder	Services	4	Female
Resp 10	CEO shareholder	Industry	9	Male

will allow us to carry out a second reading of the verbatim to ensure an objective interpretation of the information collected.

We also note that respondents expressed their opinions on culture, legitimacy, know-how and mission in accordance with CSR involvement.

After carefully rereading the statements, the verbatims were coded using the content analysis method (Bardin, 2007) in order to include them in the criteria analysed.

This content analysis of the declarations made it possible to combine the answers of different respondents to the common questions asked (Patton, 2002, p. 438), gradually revealing themes and sub-themes (Patton, 2002, p. 354, 440–441), commonly referred to as categories and sub-categories by Bardin (2007).

The analysis of the data will be based on the verbatim from the self-declarations encoded by the respondents. This guarantees a faithful consideration of the manager's perception. We have voluntarily limited the significant verbatims collected in our study to the elements expressed that are directly related to our research questions. We can consider the verbatims mobilised as a sentence or an expression testifying to a fact or a situation experienced by the respondents.

Their analysis will lead to the acceptance or not of our research proposal.

Results

French Companies' Commitment to CSR in the INSEE Study

- ***French companies' commitment to CSR***
  The descriptive study makes it possible to present the specificities linked to the different types of companies. Indeed, the commitment to customer-supplier relations increases according to the financial strength of the company. The greater the company's financial capitalization, the more it is committed to CSR and the more it feels concerned by the economic and societal benefits of its supplier

partners. The proportion of companies committed to social and environmental issues is high.

First of all, we note that 52.06%, 58.04% and 67.59% of companies with more than 19 employees and less than 250 employees respectively declared a CSR involvement. We also note that CSR and social commitment are developed in greater proportions in medium and large companies (250 and 499 employees; 80.64%) and large companies (500 employees and more; 91.44%). This is mainly due to the difficulty of perception that small and medium-sized companies have regarding the concept and definition of CSR and social involvement (Table 2.3). There is also a size effect, which is often related to organizational structure. Large companies often have a mature organisation that sets up departments dedicated to CSR. This approach is structured and responds to a clear and assertive strategic challenge. This is not the case in SMEs, where the issue is often initiated by the firm's owner or manager. Finally, the SMEs approach is often intuitive and driven by the individual values of the owner or manager. Table 2.3 shows CSR level of companies by number of employees.

- ***French companies' commitment to CSR by sector***

  Secondly, with regard to the sectoral analysis, surprisingly there is no significant difference in CSR engagement between sectors of activity. We would have expected companies in sectors related to the production and processing of natural resources to be more committed to CSR and social issues. A minimum of 97% of respondents reported being committed to CSR. The results of this study indicate that the sector to which the company is attached does not affect respondents' answers (Table 2.4). Analysis of INSEE data reveals that legislation and regulatory requirements in France have had a significant impact on companie's commitment to CSR, regardless of their sector of activity. Since the introduction of the Grenelle II law in 2010, which requires large companies to publish sustainable development reports, there has been a significant increase in CSR initiatives, as revealed by the results of the ENDD survey published in 2019 by INSEE. In addition, new regulations such as the 2019 PACTE law, which encourages the integration of social and environmental issues into corporate strategies, have consolidated this trend. Finally, companies are now more inclined to adopt sustainable practices, influenced by legislative developments that not only constrain

Table 2.3 The main factor: CSR of companies by number of employees

		Involved in CSR	
Size	8997	Yes (%)	No (%)
20–49 employees	4349	52.06	47.94
50–99 employees	1487	58.04	41.96
100–249 employees	904	67.59	32.41
250–499 employees	377	80.64	19.36
500 employees and more	1880	91.44	8.56
Total staff base	8997	64.03	35.97

Table 2.4 The second main factor: CSR of companies by sector

		Involved in CSR	
Sector	8997	Yes (%)	No (%)
Mining and quarrying, excluding AFI[a]	1889	99.36	0.64
Food and beverages	406	98.77	1.25
Energy and environment	149	99.33	0.68
Construction	985	98.78	1.23
Wholesale trade	924	98.59	1.43
Retail trade, repair and personal services	1208	9818	1.85
Transport and storage	681	98.53	1.49
Accommodation and food service activities	555	97.84	2.21
Information and communication	439	99.09	0.92
Real estate activities	163	100.00	0.00
Professional, scientific and technical activities	791	98.99	1.02
Administrative and support service activities	807	99.26	0.75
Sample	8997	98.83	1.18

[a]Agri-food industry

them, but also encourage them to incorporate CSR as a vector for competitiveness and innovation. Table 2.4 shows level of CSR of companies by sector.

Determinants of CSR Commitment in SMEs

To answer our question on the position adopted by managers, we looked at manager's perspectives of CSR and their interpretation of CSR in relation to the strategic analysis grid proposed by Courrent (2012) and Marchesnay (1993). By examining the accounts of managers actively involved in firms operating in France, we have shed light on the variables that drive the action of managers involved in CSR on the environmental, social and societal pillars.

This exploration allows us to discern the determinants that influence manager's behaviour in their commitment to CSR. The verbatims extracted from the self-disclosures of the respondents who expressed their point of view in our study enable us to identify four pivotal elements considered to be key factors in this context. These elements underlie the actions of managers and allow us to express their perceptions of responsible commitment: culture, legitimacy, know-how and mission.

- ***Culture as a driver of commitment***

When respondents associate culture with organizational goals and commitment to CSR, 40% focus on environmental issues and 40% on external social issues, while 20% focus on internal social issues.

They give their views on environmental issues:

It's a fundamental concern. (Resp 6)

Combating climate change and improving energy efficiency. Become a player in the circular economy and contribute to waste reduction. Promoting more responsible consumption and raising awareness of sustainable development issues among its stakeholders. (Resp 5)

or even:

Acting to protect the environment is, in my opinion, the first point, which will lead to better development of the company and therefore enable us to achieve the other two commitments. (Resp 10)

Finally, the respondents on this theme make the link between the company's commitment to the environment and to society:

Visibility of environmental certification. (Resp 7)

and:

As you can see, the product and the know-how are naturally oriented towards the environment. But the social and societal impact is also very much present. (Resp 3)

Societal commitment is informed by a number of criteria which show that respondents wish to add value to their societal environment in different ways by giving it meaning:

The company's culture is based on 3 key values: simplicity, responsibility and transparency. The social aspect comes first in this respect. (Resp 4)

In addition, they make the link between their culture and the expected contribution to the business relationship:

We are retailers, and what's important to us is reaching our customers. That's why our actions are aimed primarily at them. Our actions must enable them to obtain a significant benefit in terms of service and product quality, but also in terms of philosophy, because we do everything, we can to ensure that our customers have products that last as long as possible. (Resp 2)

Finally, a connection is sometimes made with a potential contribution to future generations:

Because my company is committed to children's education. (Resp 1)

Company culture is also designed to promote the well-being of company employees:

improving inclusion of the visually impaired (Resp 8)

Respondents also underlines the challenge of sharing culture. The company's vocation is to retain valuable employees and to attract talent so that they apply to join the company:

To retain and attract the talent of today and tomorrow, we need to commit to the values sought by the new generations. (Resp 9)

When we asked them to express their views on how they transpose culture into their CSR commitment, the answers below are of interest.

Firstly, the respondents indicate that they are trying to integrate stakeholders into their thinking:

By asking all stakeholders to join our company's approach. (Resp 5)

Raising the awareness of all those involved in the economic chain through appropriate communication. (Resp 6)

and:

Raising awareness and joint reflection. (Resp 10)

Openness to stakeholders and transparency are elements that are highlighted in the respondents' comments: "... A company with a mission in its articles of association, pay policy, governance, transparency in internal communications, respectful management, employment of ESATs and committed suppliers, sustainable mobility and aid, etc. (Resp 4)

Respondents also noted that culture can be valued through the value proposition offered by the firm:

Through our products and services. (Resp 8)

and:

We are working on our service offering projects, drawing on both the experience and expertise of older staff and the digital tools of younger staff, taking into account not only the economic aspects of customer sustainability, but also intangible assets. (Resp 9)

Finally, the respondents indicate that culture is also intended to capitalise on the human dimension, represented by the cultural specificities of the company:

We are a family business, we perpetuate things that may seem obvious, but which are to create a local ecosystem of supplier, customer and internal around a common goal, well-being, caring and support. Everything is done to ensure that things last as long as possible, so as to avoid false ecology, i.e. everything must be oriented towards use. We have a critical eye for our suppliers, depending on their production methods and the origin of their products. (Resp 2)

as well as:

We do this through training on the new standards and by raising awareness of individual and collective behaviour. For the past 6 months, we have also been running a training programme for all our employees based on the Climate Fresco, to raise awareness of climate change and its impact on society. This will be rolled out over 18 months. It provides a basis for sharing choices and company policy over the coming years. (Resp 3)

- ***Legitimacy as a guarantee of commitment stakeholders (or Legitimacy as a guarantee of acceptance by stakeholders)***

The legitimacy associated with the goals set with respect to the firm's environment accounts for 50% of responses on environmental issues, 30% of responses on external social commitment and 20% of responses on internal social commitment.

Respondents justify their stance in relation to the company's activity and environmental expectations:

It is important to be able to measure the impact of our activities on the environment and the aim of our actions is to set an example in terms of environmental responsibility. (Resp 5)

and

Priority due to industrial activity (Resp 6)

Aware, moreover, that the business sector to which the company belongs may, in certain situations, constitute a disadvantage:

> *Expertise, and also a handicap* (Resp 7)

Some respondents are aware that they can play a legitimate relay role in their relations with stakeholders to deal with environmental and societal issues linked to their activities:

> *We advise companies on their business strategy and sustainability, as well as on business transfers.* (Resp 8)

or even:

> *Our customers want proof of the carbon ROI, the LCA of the greenhouse, the nutritional quality, plastic residues, heavy metals, the origin of consumables, etc.* (Resp 4)

at last:

> *Due to the nature and origin of the company's creation. Founding value.* (Resp 9)

Finally, when it comes to the company's internal social issues, our response is to express our commitment to the subject:

> *In 2023, we had zero staff turnover. The search for well-being is not just a quest for meaning, but must also be materialised by facts, namely the behaviour of managers towards employees. The responsibility that is given, but also the strong autonomy that is a tradition in the company. We are also there to support the weakest, and those who encounter difficulties. The company is a social entity that is responsible for its employees. Work remains a constraint, and it's up to managers to make it as pleasant as possible by motivating those who work for the company and thanking them for the work they do.* (Resp 2)

When we ask them how they translate this legitimacy into action, here's the answer. When questioned about the implementation of the legitimacy approach, the respondents' answers tended to focus on actions of a social nature:

> *We communicate and act on concrete facts. With the Made in Elsass logo, we can communicate on the origin of our products and the proximity of our partners. Our ecosystem benefits from our actions because we have managed to lower costs and increase service quality by working with local suppliers. With upcycling, we have created a partnership with carpenters from the Kochersberg called Gentlemen designers. We renovate furniture, transforming it into bathroom furniture and improving it. This helps people to reuse products rather than throw them away. On the whole, our actions are concrete and go beyond simple recycling.* (Resp 2)

They express the recognition of their social legitimacy through actions designed to highlight their commitment and to take account of the expectations of future generations:

> *Third-party evidence, analysis reports, labels (Solar Impulse, etc.), detailed information on our approach (wiki.myfood.eu), sharing of testimonials, greenhouse applications for health, nursing homes, social integration, education, etc.* (Resp 4)

and:

We need to take account of changes in society and generational differences to be able to offer relevant advice. (Resp 9)

Respondents also stressed the need to update the firm's internal skills in order to ensure its long-term survival in a constantly changing economic environment:

Through the most demanding certifications in the profession (Resp 5)

And

Training (Resp 8)

- ***Know-how as a reason for being***

When they think of the know-how in terms of the activities developed by the organization, half of the respondents (50%) consider environmental issues to be their top priority. Less than a third feel concerned by external social issues, and 20% feel concerned by internal social issues. Then, when we asked them why they were committed in this way, the respondents said:

The environment is an essential aspect of our business as a producer of ready-mix concrete and prefabricated concrete products. The carbon footprint of our activities is 95% linked to the use of cement, so the use of low-carbon or alternative cements has a considerable impact. Our products are also used in the construction of buildings, housing, infrastructure, industry, etc. and the thermal performance of our products and their durability have a direct impact on the thermal performance of buildings and their durability (up to 100 years). This too is an environmental issue, although there is also a social aspect (for example, the construction of passive social housing, which has a positive impact on future occupants by limiting energy costs). (Resp 3)

For some respondents, a clear link has been established between environmental commitment and meeting stakeholder expectations in the context of their business:

Visibility environmental certification (Resp 7)

Communication was highlighted by three other respondents:

It's a commitment that communicates and speaks more easily to stakeholders such as customers and improves the company's footprint. It is also the best known of the three commitments. (Resp 10)

And:

Inclusion of the visually impaired in society and business (Resp 8)

But also:

A company must keep up with changes in society to remain connected to its customers/prospects of today and tomorrow, or risk losing them. (Resp 9)

A Manager cite the vital contribution of stakeholders to the firm's success:

We created the Made in Elsass logo and registered it with INPI. It's almost 15 years old. For us, local trade and local purchasing power issues have been a matter of course since our grandfather founded the company in 1958. We encourage trade with suppliers as close as possible to our businesses in Alsace. Creating innovative partnerships, particularly in the areas of upcycling and re-use. All these efforts help to create a good image for the company and make our employees proud of the work we do. Employees are also the primary beneficiaries, through their sales figures. (Resp 2)

And:

Food sovereignty necessarily means respecting our stakeholders, in particular the planet (use of resources and pollution in particular) but also our employees and suppliers, who are essential to the operation of our company, serving our families and customers. (Resp 4)

When we asked them to comment on this criterion and to describe in a few words the actions they are taking, we got two answers. When respondents talk about the link between their business activity and the dynamics of CSR commitment, they express a clear vision of their obligation to commit to mechanisms that will enable them to improve their commitment to CSR:

We express our CSR approach in the form of a company policy. This policy refers to the UN SDGs and for each of the SDGs identified, objectives and actions are defined. In addition, the company is audited annually according to the ISO 26000 standard, and we have just made an SBTi commitment to a carbon reduction trajectory. Again, this may seem like a purely environmental issue, but it has an impact on our stakeholders and our image (customers, employees, shareholders, banks, etc.). (Resp 3)

With clear operational answers to act on numerous improvement indicators:

Actions to improve life-cycle impact, choice of materials, suppliers, transport, packaging, proof of impact, measurement of water and energy use, energy ROI, reuse of rainwater and solar panels, seed sourcing, etc. This is an underlying concern in our decisions, but it is not the main issue. (Resp 4)

Respondents indicate that a large number of know-how indicators are taken into account in order to meet expectations in a changing international context: "... *It's a concern that underlies our decisions, but it's not the main issue. We need to link our intentions to the economic reality of a private, non-subsidised structure. (Resp 5)...* ".

Other respondents said that labelling, in-house skills development and other compliance measures were necessary to ensure the firm's survival:

Awareness-raising, standardisation, auditing, labelling and investment. (Resp 8)

and:

Providing services and selling technical aids. (Resp 6)

The involvement and expertise of our employees is key to the firm's success:

In an international context, I'm in contact with my customers and prospects both face-to-face and remotely, and both are useful and effective. (Resp 7)

that stakeholders outside the company:

We're part of an ALGOREL purchasing group and have launched a CSR initiative with their support. For us, CSR is also linked to innovation, so it's the equivalent of a product manager who looks after it. It's also the entire management team and the shareholders who take it on board to encourage local trade. (Resp 2)

ensure that the company creates value for all its stakeholders, both inside and outside the market.

- ***Mission as a response to environment***

When the mission criterion is approached from the point of view of the activities developed in response to the firm's environment, 70% of respondents position themselves on external societal issues when they are linked to the company's environmental activity in a CSR context. In addition, 30% take a position on environmental issues.

Respondents position themselves on societal issues related to their missions by expressing their opinion as follows:

> *The mission and the goal to achieve, an ideal towards which we want to have more Construction of buildings, with a use of materials more in line with what is needed to make it last. We need to find ways of enabling as many people as possible to live in and renovate homes with consistent budgets and quality materials. Our slogan is to transform our living spaces into dream places.* (Resp 2)

A part of their mission is to keep in mind a response expected by their stakeholders:

> *Priority due to industrial activity* (Resp 7)

And:

> *Creation of a dedicated product. (Resp 8*)

Finally, the last respondent gave the same answer as for the legitimacy criterion:

> *Same as for legitimacy … Due to the nature and origin of the company's creation. Founding value.* (Resp 9)

One respondent commented on the societal value of his know-how, saying:

> *The company's mission is to give all citizens back the opportunity to produce their own food and regain their autonomy. It's a social mission to promote food sovereignty in the home and combat inflation.* (Resp 4)

When we asked them to comment on this theme mission as part of their CSR commitment, here was their response.

The respondent indicates that they are implementing approaches that are designed to enable them to create value for their customers:

> *Product sourcing, supplier selection, product sustainability, product origin: we communicate this information in our catalogues and through our various media.* (Resp 2)

But also, for their stakeholders in the broadest sense:

> *Innovative greenhouses to feed a family all year round, in a context where time, knowledge and space are in short supply.* (Resp 4)

By encouraging the commitment of everyone involved:

> *Raising awareness.* (Resp 7)

One respondent also repeating an element indicated in the legitimacy section:

> *Same as for legitimacy … We advise companies on their business strategy and sustainability, as well as on business transfers.* (Resp 8)

Discussion

The commitment of managers to Corporate Social Responsibility (CSR) is a multifaceted phenomenon influenced by several determinants. Furthermore, the results of the study confirm the major role played by SME managers, which in their organisational positioning are considered to be top managers in the sense of Swanson (2008). In our study, we explore how culture, among other factors, shapes managerial behavior in their commitment to CSR, focusing on the environmental, social, and societal pillars. So, the four criteria are identified by the respondents as being levers for commitment to the three pillars of (Gherib et al., 2009). These results also confirm that in most of the firms responding, CSR is discretionary and sometimes CSR practices are not very formalised (Lapointe & Gendron, 2005; Berger-Douce, 2006).

- ***Culture an essential prerequisite for responsible commitment***

  The exploration of managerial perspectives on CSR within French firms reveals that culture is a fundamental driver of CSR commitment, influencing actions across environmental, social, and societal pillars. Cultural values embedded within organizations guide their commitment to CSR, ensuring that environmental stewardship, social responsibility, and societal impact are integrated into business practices. By fostering a culture of responsibility, inclusivity, and sustainability, companies can effectively navigate the complexities of CSR and contribute to a more sustainable and equitable future. These findings underscore the importance of aligning organizational culture with CSR objectives, highlighting how deeply ingrained values and norms shape managerial behavior and drive responsible corporate actions.

- ***Legitimacy as a corporate recognition tool***

  Legitimacy is a critical criterion in the commitment to CSR, influencing how SMEs managers align their actions with societal and market expectations. Managerial perspectives highlight the importance of legitimacy in driving CSR initiatives across environmental, social, and societal pillars. By fostering a supportive work environment, demonstrating transparency in environmental actions, and addressing societal challenges, companies can enhance their legitimacy and strengthen their CSR commitments.

  These insights underscore the need for SMEs managers to continuously align their CSR strategies with societal norms and expectations, ensuring that their actions are perceived as legitimate and impactful. This alignment not only enhances the firm's reputation but also contributes to long-term sustainability and stakeholder trust.

- ***Know-how as a source of differentiation and innovation***
  Corporate Social Responsibility (CSR) is increasingly recognized as integral to the strategic framework of contemporary businesses. This discussion aims to elucidate the role of know-how as a "reason for being" in CSR, focusing on how organisational capabilities influence engagement in environmental, social, and societal pillars. Insights are drawn from responses of SMEs managers, highlight-

ing the interconnection between business activities and CSR. This approach not only enhances the legitimacy of their CSR efforts but also contributes to sustainable business practices and innovation that benefit all stakeholders (Berger-Douce, 2006).

- ***Mission as a lighthouse for goal setting***

The concept of mission is crucial for the strategic alignment of Corporate Social Responsibility (CSR) initiatives within an organization's operational framework. This analysis explores how SMEs managers incorporate their mission into CSR activities, with a particular focus on environmental, social, and societal aspects. The study draws on responses from SME managers, highlighting the harmonization between corporate mission and CSR commitments.

Embedding the mission into CSR efforts is vital to ensure alignment with the SMEs strategic objectives. By integrating environmental, social, and societal goals into their core missions, firms can create a cohesive and impactful CSR strategy that resonates with stakeholders and promotes long-term sustainability.

Conclusion

This academic exploration of Corporate Social Responsibility (CSR) commitment within French companies and particularly in SMEs underscores the critical role of organizational culture, legitimacy, know-how and mission in shaping effective CSR strategies. Our qualitative study reveals that these elements are deeply intertwined, collectively guiding managerial actions and enhancing the organization's overall commitment to CSR across environmental, social, and societal pillars. We take up each of the four criteria analysed to develop their contribution to the CSR involvement of SMEs managers.

Cultural values as CSR catalysts: organizational culture, encompassing values and norms, serves as a fundamental driver of CSR commitment. The study illustrates how cultural emphasis on environmental stewardship, social responsibility, and societal impact influences managerial behavior and CSR initiatives. Cultural alignment fosters a supportive environment where CSR becomes a natural extension of daily business practices.

Legitimacy and CSR engagement: legitimacy, rooted in societal norms and market expectations, is pivotal in sustaining CSR efforts. Companies enhance their legitimacy by aligning their actions with societal values, thus meeting stakeholder expectations. This alignment not only bolsters the firm's reputation but also ensures long-term sustainability through transparent and accountable practices.

Know-how and organizational capabilities: CSR initiatives are most effective when they are integrated into the core business activities and supported by robust organizational capabilities. From sustainable product development to inclusive network access, companies demonstrate that strategic CSR commitments can drive both business success and positive societal impact.

Mission as a strategic driver: the alignment of CSR initiatives with the company's mission ensures that CSR efforts are not peripheral activities but integral to the organization's core values and strategic objectives. By embedding sustainability, social responsibility, and technological equity into their mission statements, companies can foster a coherent CSR strategy that enhances legitimacy and stakeholder trust.

These findings underscore the importance of a holistic approach to CSR, where culture, legitimacy, know-how and mission are seamlessly integrated to create a coherent and impactful strategy. SMEs Managers play a crucial role in translating these elements into actionable initiatives that resonate with stakeholders and contribute to a sustainable and equitable future.

Limitations and Future Research Directions

Despite the comprehensive insights provided by this study, several limitations must be acknowledged. First, the study is based on qualitative data from a specific geographical and cultural context, which may limit the generalizability of the findings. Future research could expand the scope to include diverse cultural and organizational contexts to validate and extend these insights.

Second, while the study highlights the importance of alignment between culture, legitimacy, know-how and mission it does not deeply explore the dynamic interactions and potential conflicts between these elements. Further research could investigate these interactions in greater detail, examining how organizations navigate potential trade-offs and synergies.

Third, the study primarily focuses on managerial perspectives, which, while valuable, may not fully capture the experiences and insights of other stakeholders such as employees, customers, and community members. Incorporating a broader range of stakeholder perspectives could provide a more comprehensive understanding of CSR commitment and its impacts.

Fourth, using secondary data requires the implementation of careful preparatory work to be able to collect the data adapted to the research object. Thus, the sources, the methodology, the database used, their structuring, the recency or even the sample used must be the subject of particular attention to guarantee a robust contribution to the research work undertaken (Snow & Thomas, 1994; Zikopoulos et al. 2012). In addition, the secondary data available do not always fully respond to research work deemed to be innovative in the academic field. Researchers must therefore be careful to mobilize only the information necessary for their subject of study. The declarations and the specificities linked to the respondents can constitute notable limits to the exploitation of this type of data (Venkatraman & Ramanujam, 1985).

In the context of self-reported data, it is imperative to acknowledge the presence of inherent biases, as respondents may occasionally exhibit a tendency to accentuate their declarations. The interpretation of a given subject may undergo diverse understandings among respondents. In this regard, we introduce the concept of nation, territory, industry, sector of activity, or market, emphasizing the need for a nuanced interpretation that accounts for these aforementioned indicators (Fratocchi et al. 2014). Such considerations may significantly alter the obtained results in comparison to the anticipated outcomes of the initially conducted research. Hence, meticulous efforts are imperative to appropriately format the data, filtering out information that is extraneous to the study's objectives. This ensures the usability of the data for conducting an optimal analysis.

As our study adopts an exploratory approach in its qualitative part, we propose an expansion of this research in the future by leveraging resources aligned with CSR involvement within the model proposed by Courrent (2012) and Marchesnay (1993). The utilization of this model would enable the identification of strategies employed by SME's managers to foster CSR involvement. The model shows significant promise for SME managers aiming to articulate and implement practices to navigate the challenges posed by CSR involvement in a more streamlined manner.

Future research could also delve into the specific mechanisms through which organizations can enhance the alignment between their mission and CSR initiatives. This includes exploring the role of organizational capabilities, stakeholder engagement, and strategic communication in fostering effective CSR strategies. Additionally, longitudinal studies could provide insights into how CSR commitments evolve over time and their long-term impacts on organizational performance and societal well-being.

By addressing these limitations and pursuing these research directions, scholars and practitioners can deepen their understanding of CSR and develop more effective strategies for integrating social responsibility into the core fabric of organizational operations.

Data Availability

Considering the confidential nature of the investigation, the anonymized data that underpins the conclusions of this study can be obtained from the corresponding author upon a reasonable request.

Information:

"This work benefited from a government grant managed by the National Research Agency as part of the Future Investments program bearing the reference ANR-10-EQPX-17 (Secure Data Access Center-CASD)."

References

Aguinis, H., & Glavas, A. (2012). What we know and don't know about corporate social responsibility: A review and research agenda. *Journal of Management, 38*(4), 932–968.

Bansal, P., & Roth, K. (2000). Why companies go green: A model of ecological responsiveness. *Academy of Management Journal, 43*(4), 717–736.

Bardin, L. (2007). *L'analyse de contenu*. Presses universitaires de France.

Berger-Douce, S. (2006). L'appropriation de la RSE par les PME: la démarche collective au service de l'engagement environnemental. *Bulletin Oeconomia Humana: Bulletin de La Chaire Économie et Humanisme, 4*(11), 24.

Boakye, D. J., Tingbani, I., Ahinful, G., Damoah, I., & Tauringana, V. (2020). Sustainable environmental practices and financial performance: Evidence from listed small and medium-sized enterprise in the United Kingdom. *Business Strategy and the Environment, 29*(6), 2583–2602.

Bowen, H. R. (2013). *Social responsibilities of the businessman*. University of Iowa Press.

Carroll, A. B. (1979). Corporate performance. *The Academy of Management Review, 4*, 497–505.

Carroll, A. B. (2016). Carroll's pyramid of CSR: Taking another look. *International Journal of Corporate Social Responsibility, 1*, 1–8.

Carroll, A. B., & Buchholtz, A. K. (2015). Corporate citizenship: Social responsibility, responsiveness, and performance. *Classics of Organization Theory, 439*.

Chamberlin, E. H. (1953). The product as an economic variable. *The Quarterly Journal of Economics, 67*(1), 1–29.

Chassé, S., & Courrent, J. (2018). Linking owner–managers' personal sustainability behaviors and corporate practices in SMEs: The moderating roles of perceived advantages and environmental hostility. *Business Ethics: A European Review, 27*(2), 127–143.

Clarkson, M. E. (1995). A stakeholder framework for analyzing and evaluating corporate social performance. *Academy of Management Review, 20*(1), 92–117.

Courrent, J. (2012). *RSE et développement durable en PME, Bruxelles, de Boeck, Coll.* Petites Entreprises et Entrepreneuriat.

Crossley, R. M., Elmagrhi, M. H., & Ntim, C. G. (2021). Sustainability and legitimacy theory: The case of sustainable social and environmental practices of small and medium-sized enterprises. *Business Strategy and the Environment, 30*(8), 3740–3762.

Dahlsrud, A. (2008). How corporate social responsibility is defined: An analysis of 37 definitions. *Corporate Social Responsibility and Environmental Management, 15*(1), 1–13.

Davis, K. (1960). Can business afford to ignore social responsibilities? *California Management Review, 2*(3), 70–76.

Davis, K. (1973). The case for and against business assumption of social responsibilities. *Academy of Management Journal, 16*(2), 312–322.

Dyck, B., & Silvestre, B. S. (2018). Enhancing socio-ecological value creation through sustainable innovation 2.0: Moving away from maximizing financial value capture. *Journal of Cleaner Production, 171*, 1593–1604.

Elkington, J. (1998). Partnerships from cannibals with forks: The triple bottom line of 21st-century business. *Environmental Quality Management, 8*(1), 37–51.

Ellram, L. M., & Tate, W. L. (2016). The use of secondary data in purchasing and supply management (P/SM) research. *Journal of Purchasing and Supply Management, 22*(4), 250–254.

Fel, F. (2011). Maturité des démarches RSE et achats durables. *Revue Sciences de Gestion, 84*.

Feller, J. (1977). L'Analyse du contenu, de L. Bardin. *Communication and Langages, 35*(1), 123–124.

Forza, C. (2002). Survey research in operations management: A process-based perspective. *International Journal of Operations and Production Management, 22*(2), 152–194.

Fratocchi, L., Di Mauro, C., Barbieri, P., Nassimbeni, G., & Zanoni, A. (2014). When manufacturing moves back: Concepts and questions. *Journal of Purchasing and Supply Management, 20*(1), 54–59.

Frederick, W. C. (2018). Corporate social responsibility: From founders to Millennials. In *Corporate social responsibility* (pp. 3–38). Emerald Publishing Limited.

Freeman, R. E. (2010). *Strategic management: A stakeholder approach*. Cambridge University Press.

Friedman, M. (1953). The methodology of positive economics.

Friedman, M. (1970). A Friedman doctrine: The social responsibility of business is to increase its profits. *The New York Times Magazine, 13*(1970), 32–33.

Gherib, J. B. B., Spence, M., & Biwolé, V. O. (2009). Développement durable et PME dans les pays émergents: Entre proactivité, opportunisme et compromis. *Journal of Small Business and Entrepreneurship, 22*(3), 355–375.

Graafland, J. (2018). Does corporate social responsibility put reputation at risk by inviting activist targeting? An empirical test among European SMEs. *Corporate Social Responsibility and Environmental Management, 25*(1), 1–13.

Hammann, E., Habisch, A., & Pechlaner, H. (2009). Values that create value: Socially responsible business practices in SMEs–empirical evidence from German companies. *Business Ethics: A European Review, 18*(1), 37–51.

Hill, C. W., & Jones, T. M. (1992). Stakeholder-agency theory. *Journal of Management Studies, 29*(2), 131–154.

INSEE, enquête développement durable ENDD 2016, « La responsabilité sociétale des entreprises: une démarche déjà répandue », ***INSEE Focus*** n° 155, mai 2019, https://www.insee.fr/fr/statistiques/4134329

Jamali, D., Zanhour, M., & Keshishian, T. (2009). Peculiar strengths and relational attributes of SMEs in the context of CSR. *Journal of Business Ethics, 87*, 355–377.

Jenkins, H. (2004). A critique of conventional CSR theory: An SME perspective. *Journal of General Management, 29*(4), 37–57.

Jones, T. M. (1980). Corporate social responsibility revisited, redefined. *California Management Review, 22*(3), 59–67.

Jones, M. T. (1999). The institutional determinants of social responsibility. *Journal of Business Ethics, 20*, 163–179.

Juholin, E. (2004). For business or the good of all? A Finnish approach to corporate social responsibility. *Corporate Governance: The International Journal of Business in Society, 4*(3), 20–31.

Lapointe, A., & Gendron, C. (2005). *La responsabilité sociale d'entreprise dans la PME: Option marginale ou enjeu vital?* École des sciences de la gestion, Université du Québec à Montréal.

Levratto, N. (2009). *Les PME: Définition, rôle économique et politiques publiques*. De Boeck Supérieur.

Lin, Y.-C., Padliansyah, R., & Lin, T.-C. (2020). The relationship and development trend of corporate social responsibility (CSR) literature: Utilizing bibliographic coupling analysis and social network analysis. *Management Decision, 58*(4), 601–624.

Lozano, R., Carpenter, A., & Huisingh, D. (2015). A review of 'theories of the firm' and their contributions to Corporate Sustainability. *Journal of Cleaner Production, 106*, 430–442.

Lu, H. E., Potter, A., Sanchez Rodrigues, V., & Walker, H. (2018). Exploring sustainable supply chain management: A social network perspective. *Supply Chain Management: An International Journal, 23*(4), 257–277.

Madsen, H., & Ulhøi, J. P. (1996). Environmental management in Danish manufacturing companies: Attitudes and actions. *Business Strategy and the Environment, 5*(1), 22–29.

Malhotra, M. K., & Grover, V. (1998). An assessment of survey research in POM: From constructs to theory. *Journal of Operations Management, 16*(4), 407–425.

Marchesnay, M. (1993). *Management stratégique*. Eyrolles Paris.

McGuire, J. W. (1969). The changing nature of business responsibilities. *The Journal of Risk and Insurance, 36*(1), 55–61.

Paradas, A. (2007). Le dirigeant comme levier de la RSE en TPE: approche exploratoire basée sur l'utilisation de récits et d'une cartographie cognitive. *Revue Internationale PME, 20*(3), 43–67.

Park-Poaps, H., & Rees, K. (2010). Stakeholder forces of socially responsible supply chain management orientation. *Journal of Business Ethics, 92*, 305–322.

Patton, M. Q. (2002). *Qualitative research & evaluation methods*. Sage.

Porter, M. E., & van der Linde, C. (1995a). Toward a new conception of the environment-competitiveness relationship. *Journal of Economic Perspectives, 9*(4), 97–118.

Porter, M., & Van der Linde, C. (1995b). Green and competitive: Ending the stalemate. *The Dynamics of the Eco-Efficient Economy: Environmental Regulation and Competitive Advantage, 33*, 120–134.

Rockmann, K. W., & Vough, H. C. (2023). Using quotes to present claims: Practices for the writing stages of qualitative research. *Organizational Research Methods, 10944281231210558.*

Sandin, G., & Peters, G. M. (2018). Environmental impact of textile reuse and recycling–A review. *Journal of Cleaner Production, 184*, 353–365.

Snow, C. C., & Thomas, J. B. (1994). Field research methods in strategic management: Contributions to theory building and testing. *Journal of Management Studies, 31*(4), 457–480.

Spence, L. J. (1999). Does size matter? The state of the art in small business ethics. *Business Ethics: A European Review, 8*(3), 163–174.

Spence, L. J., Rutherfoord, R., & Blackburn, R. A. (1998). *Small business and environmental issues in the UK and the Netherlands: A literature review and research agenda*. Small Business Research Centre.

Swanson, D. L. (1995). Addressing a theoretical problem by reorienting the corporate social performance model. *Academy of Management Review, 20*(1), 43–64.

Swanson, D. L. (2008). *Top managers as drivers for corporate social responsibility.*

Thiétart, R.-A. (2014). *Méthodes de recherche en management-4ème édition*. Dunod.

Vandchali, H. R., Cahoon, S., & Chen, S.-L. (2021). The impact of supply chain network structure on relationship management strategies: An empirical investigation of sustainability practices in retailers. *Sustainable Production and Consumption, 28*, 281–299.

Venkatraman, N., & Ramanujam, V. (1985). *Construct validation of business economic performance measures: A structural equation modeling approach*. BEBR Faculty Working Paper; No. 1148.

Wacker, J. G. (1998). A definition of theory: Research guidelines for different theory-building research methods in operations management. *Journal of Operations Management, 16*(4), 361–385.

Wang, C., Zhang, Q., & Zhang, W. (2020). Corporate social responsibility, Green supply chain management and firm performance: The moderating role of big-data analytics capability. *Research in Transportation Business & Management, 37*, 100557.

Wartick, S. L., & Cochran, P. L. (1985). The evolution of the corporate social performance model. *Academy of Management Review, 10*(4), 758–769.

Wood, D. J. (1991). Social issues in management: Theory and research in corporate social performance. *Journal of Management, 17*(2), 383–406.

Wood, D. J., Mitchell, R. K., Agle, B. R., & Bryan, L. M. (2021). Stakeholder identification and salience after 20 years: Progress, problems, and prospects. *Business and Society, 60*(1), 196–245.

Yawar, S. A., & Seuring, S. (2018). The role of supplier development in managing social and societal issues in supply chains. *Journal of Cleaner Production, 182*, 227–237.

Yu, W., & Ramanathan, R. (2015). An empirical examination of stakeholder pressures, green operations practices and environmental performance. *International Journal of Production Research, 53*(21), 6390–6407.

Zhu, Q., Zou, F., & Zhang, P. (2019). The role of innovation for performance improvement through corporate social responsibility practices among small and medium-sized suppliers in China. *Corporate Social Responsibility and Environmental Management, 26*(2), 341–350.

Zikopoulos, P., Eaton, C., deRoos, D., Deutsch, T. & Lapis, G. (2012). *Understanding big data: Analytics for enterprise class hadoop and streaming data*. New York, McGraw-Hill Osborne Media.

Stéphano Vacher holds a PhD in Management Sciences obtained at University of Clermont Auvergne in 2019 and is Associate Professor at EM Strasbourg-UNISTRA since 2021. Is the Head of the Master European Digital & Sustainable Business Trinational, and the Head of Strategic Axis on Bachelor International Affairs. He is a member of Humanis (Human In Society), his main areas of research interest are innovation, Corporate Social Responsibility, Corporate Strategy, Governance, Leadership, and Value Creation. His current research focuses on technical innovation, the firm's CSR involvement and ecological transition, and a new focus concerning sustainable supply chain relationships and eco-innovation. He is also a member of the AIMS, RIRL, RIODD and CSR & Value Chair. He has more than 15 years of experience in Marketing Strategy, Sales, Business Consulting and as a Sales engineer. Dr. VACHER's pedagogy is action-oriented with the use of concrete cases from partnerships in the field with market players and the mobilization of case studies produced regularly to meet the expectations of learners.

Chapter 3
Core Values of Sustainable Innovation in Indonesian SMEs

Giorgio Gatica Singgih and Melia Famiola

Abstract In response to growing environmental concerns and global sustainability goals, businesses are increasingly adopting more sustainable practices, a move strongly supported by stakeholders worldwide. However, for companies in Indonesia, a developing country, balancing the need to consider environmental impacts with pursuing profits poses a challenge. Small and Medium Enterprises (SMEs) are vital to Indonesia's economy and play a crucial role in this context. To address this challenge, this paper aims to examine the sustainability values and practices of certified B Corps SMEs in Indonesia, using secondary data from B Corp. Through thematic analysis, a better understanding of the environmental, social, and governance characteristics of Indonesian SMEs' practices can be gained.

This research provides a comprehensive understanding of sustainability innovation within SMEs. It also offers valuable insights for stakeholders and policymakers seeking to enhance environmental and social impacts in the Small and Medium Enterprises sector, particularly for their practices in Indonesia.

Keywords Sustainable practices in SMEs · B Corps Certification · Sustainability innovation · Indonesia

Introduction

SMEs are instrumental in job creation, inclusive economic growth, and fostering a fair and just society. Studies indicate that SMEs absorb up to 97% of Indonesia's workforce and contribute over 60% of the national GDP (Permatasari et al., 2023; Famiola, 2020). Given this significant impact, SMEs can potentially contribute to poverty alleviation. By prioritizing sustainability, these enterprises can ensure that economic growth is environmentally and socially responsible. The benefits of

G. G. Singgih (✉) · M. Famiola
School of Business Management, Institut Teknologi Bandung, Bandung, Indonesia
e-mail: giorgio_gatica@sbm-itb.ac.id; melia.famiola@sbm-itb.ac.id

A. H. Verkuil et al. (eds.), *Core Values and Decision-Making for Sustainable Business*, Sustainable Business Development,
https://doi.org/10.1007/978-3-031-78361-6_3

sustainable practices, such as improved working conditions and potentially higher wages, can further enhance the livelihoods of those employed by SMEs, highlighting the positive outcomes of sustainability initiatives.

However, SMEs face distinct challenges in implementing sustainable practices. One significant barrier is the widespread lack of awareness and knowledge among SME owners regarding the environmental implications of their business activities (Yuliani & Soetjipto, 2019). This lack of understanding extends to limited knowledge of environmental laws and regulations, complicating compliance efforts. Economic constraints also pose significant obstacles, particularly for small and micro-enterprises with limited funding and resources. These financial limitations make implementing sustainable practices economically burdensome, hindering progress toward sustainability (Setyaningsih et al., 2024). Addressing these challenges is crucial for policymakers and stakeholders to facilitate the adoption of sustainable practices among SMEs.

In addition to economic challenges, regulatory barriers impede the advancement of sustainability initiatives among Indonesian SMEs. Insufficient enforcement of environmental regulations, such as Law No. 18/2008 on Waste Management, highlights the need for more robust incentives or sanctions to ensure compliance with Environmental, Social, and Governance (ESG) frameworks (Haruuma, 2023). Furthermore, socio-cultural factors add complexity to sustainability adoption, as a lack of customer knowledge and understanding regarding sustainability hinders the widespread adoption of eco-friendly practices (Gunawan et al., 2022).

Despite these challenges, many Indonesian SMEs strive to make sustainability a core aspect of their business. These enterprises are often led by founders who embrace a sustainability mindset, influenced by their educational background and traditional values, which serve as the foundation for their business strategy (Anggarini, 2020; Famiola, 2020). Some of these businesses have taken steps to obtain B Corp certification, a standard developed by a non-profit organization leading the global push towards a more inclusive, equitable, and regenerative economy. B Corp certification demonstrates a company's commitment to implementing Sustainable Development Goals (SDGs), promoting environmental stewardship, good governance, and creating social impact.

This study aims to assess the sustainability innovation practices of Indonesian SMEs, focusing on those that have received B Corp certification. By analyzing the innovation characteristics in these SMEs, we can identify best practices that can serve as a model for other SMEs seeking to prioritize sustainability as a critical value for their growth.

The Perspective of Sustainability Innovation

Sustainability, as a concept, represents various contexts and activities and is an ongoing journey that will increasingly shape the orientation of business innovation (Adams et al., 2016). Adams et al. (2016) define three stages in a business's journey toward becoming a sustainable corporation: operational optimization, organizational transformation, and system building.

1. **Operational optimization** focuses on enhancing internal business practices with incremental changes to increase efficiency and reduce negative environmental impacts, often for compliance. Companies at this stage typically view sustainability initiatives as a means to mitigate adverse effects rather than as opportunities for innovation.
2. **Organizational transformation** marks a shift in mindset, where companies begin to see sustainability as an opportunity rather than a mere compliance obligation. At this stage, businesses innovate by developing new products or services aligned with their sustainability agenda, demonstrating a more profound commitment to sustainability beyond minimizing negative impacts.
3. **System building** represents the highest commitment to sustainability, where companies innovate to create and support a new and improved social system. This stage involves a radical shift in philosophy, with businesses broadening their focus beyond individual goals to consider broader societal objectives. Sustainability at this level is seen as a global attribute, requiring extensive relationships and collaboration to achieve collective impact, emphasizing the interconnectedness of business and societal goals.

Five key activity categories—strategy, process, learning, linkages, and innovative organization—drive the three stages of transformation.

1. **Strategy** involves integrating sustainability perspectives into core business strategies and decision-making processes. Hallstedt et al. (2013) highlight the importance of embedding a strategic sustainability perspective into product innovation, ensuring sustainability becomes a fundamental component of business planning and operations.
2. **Process** focuses on the practical implementation of sustainability within operations. Ceschin (2013) emphasizes the need to address corporate, cultural, and regulatory barriers to implement eco-efficient product-service system innovations effectively. Effective process management is crucial for translating sustainability strategies into tangible actions and outcomes.
3. **Learning** entails continuous improvement and adaptation based on new knowledge and experiences. Assink (2006) and Teece (2007) underscore the necessity for firms to develop the capacity to adapt, integrate, and reconfigure organizational skills, resources, and functional competencies in response to sustainability challenges. This ongoing learning process is essential for progressing from operational optimization to more transformative practices.
4. **Linkages** involve collaboration and integration with external stakeholders, including NGOs, regulatory bodies, and other firms. Organizations like the World Business Council for Sustainable Development (WBCSD) and the World Wide Fund for Nature (WWF) are critical in bridging the gap between science and business, promoting corporate sustainability rooted in ecological principles.

Effective linkages enable firms to leverage external expertise and resources, enhancing their sustainability efforts through collective action.

5. **Innovative organization** refers to fundamentally rethinking and restructuring business models to align with sustainability goals. Bocken et al. (2014) propose sustainable business model archetypes, such as maximizing resource productivity, creating value from waste, and substituting non-renewables with renewables. The shift towards circular economy models reflects a move away from linear industrial paradigms, promoting systemic innovation and long-term sustainability.

These five categories collectively provide a comprehensive framework for understanding the diverse activities involved in sustainability innovation. By focusing on strategy, process, learning, linkages, and innovative organization, firms can navigate the complex sustainability landscape and drive meaningful, long-term change. This approach aligns with the need for a dynamic and adaptable model that reflects contemporary practices and the evolving nature of sustainability challenges.

The three stages of transformation have distinct characteristics associated with these five activities, as outlined in Table 3.1.

Using the classifiers outlined above, this study examines the characteristics of sustainable innovation in small and medium-sized enterprises (SMEs) in Indonesia. The goal is to identify practical approaches to developing sustainable SMEs, particularly in Indonesia, that can also be implemented by SMEs in other developing countries.

Table 3.1 The stage of Sustainable Innovation and the activities

	Operational optimization (Code OO)	Organizational transformation (Code OT)	Systems building (Code SB)
Strategy	Regulation compliance and efficiency gains	Embedding sustainability	Wide collaborations and systems solutions
Process	Internal and incremental innovation	Adopting new values, platforms, and ideation practices	Collaborative processes with diverse stakeholders
Learning	Knowledge management and access	Engagement with key stakeholders	Developing ambidextrous skills
Linkage	Recruiting external domain experts	Shift in focus to collaborations with stakeholders	Involving stakeholders in problem diagnosis and change identification
Innovative organization	Exploiting existing innovation capabilities	Embedding culture of innovation	Consideration of new business paradigms

B-Corporations Certified

This study uses B-Corp Certified SMEs as a primary benchmark to substantiate the social and environmental sustainability impact of SMEs. The B Impact Assessment (BIA) serves as a comprehensive tool for managing and evaluating impact, divided into five distinct domains: Governance, Workers, Community, Environment, and Customers. Each domain encompasses both Operational Impact (white) and Impact Business Model (colored) components (see Fig. 3.1).

In the **Governance** domain, sub-dimensions include Mission and engagement and Ethics and transparency as operational impacts, with Mission Locked as the impact business model. The worker's domain covers financial security, health, wellness, safety, career development, engagement, and satisfaction as operational impacts, and workforce development and worker-owned as impact business models. The **Community** domain includes Diversity, Equity and inclusion, Economic Impact, Civic Engagement and giving, and Supply Chain Management as operational impacts, along with impact business models such as Supply Chain Poverty Alleviation and Local Economic Development. The **Environment** domain focuses on Environmental Management, Air and climate, Water, and Land and life for

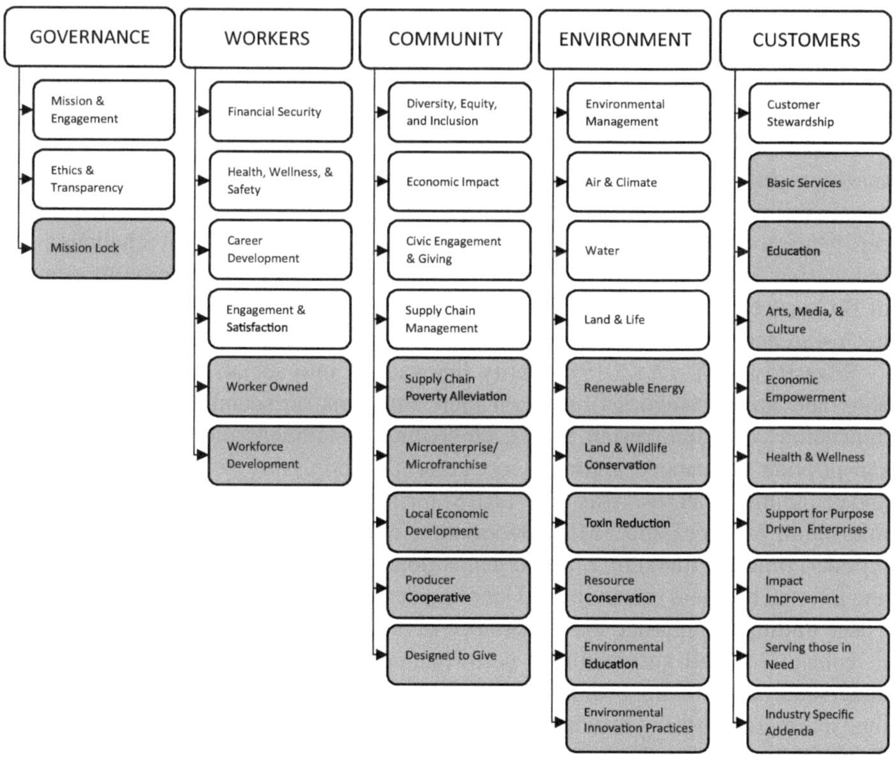

Fig. 3.1 B impact assessment. Source: B Lab (2020a)

operational impacts, with impact business models like Land/Wildlife Conservation and Renewable or Cleaner-Burning Energy. Lastly, the **Customers** domain includes Customer Stewardship as an operational impact and several impact business models, such as Basic Service, Education, and Health & Wellness Improvement, emphasizing positive customer outcomes and privacy protection (B Lab, 2020c).

Each domain evaluates the effects of a company's operations and business model on stakeholders, distinguishing between operational impact and impact business models. The BIA employs a scoring mechanism for comparability and tracking improvement opportunities over time. The scoring process is objective, normalized, and materiality-based, enabling all companies to earn the same total points in the assessment. Points are allocated based on the difficulty of implementing practices and the directness of indicators in assessing positive impacts. Weighted questions reflect a company's positive impact based on policies, practices, outputs, and outcomes, while unweighted metric questions provide additional context to enhance self-reporting accuracy.

The BIA offers a comprehensive framework that enables a holistic evaluation of a company's impact across multiple sustainability domains. This structured approach allows for consistent benchmarking, providing a standardized mechanism to compare and track the performance of different companies. Additionally, the BIA is designed to focus on continuous improvement, guiding companies to identify and enhance areas of their sustainability practices over time.

However, the BIA also presents certain challenges. Its complexity can be daunting, particularly for smaller SMEs with limited resources, as the detailed and weighted questions often require extensive data collection and analysis. Moreover, sustainability is frequently perceived as a cost rather than an investment in Indonesia. This perception is compounded by the need for SMEs targeting the local, price-sensitive market to adopt new market penetration strategies to introduce sustainability-focused products. The financial burden is further exacerbated by the certification fees, which are paid in U.S. dollars—a significant challenge for Indonesian SMEs that primarily earn revenue in Rupiah. Furthermore, the concept of B Corp certification is not yet widely recognized in Indonesia, resulting in limited awareness and low adoption among local businesses.

Silvestre and Ţîrcă (2019) classify innovations into social, traditional, sustainable, and green categories based on their emphasis on social and environmental dimensions. For Indonesian SMEs, embracing sustainable and green innovations can provide a balanced approach, concurrently addressing social, environmental, and economic concerns. Asad et al. (2018) further highlight the relationship between innovation practices and SME performance, underscoring the favorable impact of product, process, marketing, and organizational innovation on financial, customer, and internal business processes and learning and growth dimensions of performance. These findings emphasize the importance of adopting suitable innovation strategies to enhance overall sustainability performance. Finally, Muangmee et al. (2021) underscore the influence of Green Entrepreneurial Orientation (GEO) on green innovation and firms' ensuing social, economic, and environmental performance. Their findings affirm the positive correlation between GEO, green innovation, and sustainable performance, highlighting the need for proactive green strategies among SMEs. Consequently, SMEs must assess their current standing and devise measures to incentivize further adoption of sustainable development initiatives.

Indonesian B Corp SMEs

In Indonesia, 24 companies hold B-Corp certification, but only 10 are classified as small and medium-sized enterprises (SMEs). This classification is corroborated by media sources that recognize these companies as part of the SME community, recipients of grants, and other relevant acknowledgments. The selected SMEs operate in the wholesale/retail, manufacturing, and service provider sectors with significant environmental footprints. These sectors are particularly relevant for this study due to their potential for environmental harm and the necessity of managing their impact effectively. Table 3.2 provides a short description of all the companies that become our object of study.

Table 3.2 Short description of studied SMEs

No	Company name	Business description	Area of impact	Products	Location
1	Bhineka Rahsa Nusantara	An SME empowers local women as artisans to develop herbal products that have become local wisdom across Indonesia and are marketed nationwide.	Cultural Preservation and Economic Empowerment	Herbal products developed and marketed by local women artisans, rooted in traditional Indonesian wisdom.	Bandung
2	Alert Green LTD	SMEs in Yogyakarta, East Java, Indonesia, are empowering local farmers to produce coconut sugar products from organic materials.	Environmental Sustainability and Organic Farming	Coconut sugar produced from organic materials by local farmers in Yogyakarta, East Java.	Hargotirto Village, Yogyakarta
3	Instellar Indonesia	Engaged in consulting to help start-up companies become impact-oriented companies. Engaged in consulting to help start-up companies become impact-oriented companies.	Social Impact Consulting and Startup Support	Consulting services to help startups become impact-oriented and integrate sustainability into their core operations.	Jakarta

(continued)

Table 3.2 (continued)

No	Company name	Business description	Area of impact	Products	Location
4	Kmana	A company operating in the fashion industry, especially luxury leather bags, by empowering Indonesia local artisans	Artisan Empowerment and Sustainable Fashion	Luxury leather bags created by Indonesian local artisans, combining high-quality craftsmanship with sustainable practices.	Seminyak, Bali
5	Lusana	An Australian lifestyle brand offering sustainable, contemporary resort wear and handwoven accessories that celebrate the female form with comfort, elegance, and timeless.	Sustainable Fashion and Resort Wear	Sustainable resort wear and handwoven accessories designed to celebrate elegance and timelessness.	Bali
6	Gree Energy Limited	A company empowers food processors in emerging countries to deploy biogas solutions that treat industrial wastewater and cut methane emissions.	Environmental Sustainability and Waste Management	Biogas solutions for treating industrial wastewater and reducing methane emissions.	Jakarta
7	SukkhaCitta	An SME specializing in regenerative fashion, the farm-to-closet brand fosters economic empowerment for women artisans and smallholder farmers in Indonesian villages while preserving indigenous culture and regenerating the planet	Regenerative Fashion and Economic Empowerment	Fashion products developed using a farm-to-closet model, focusing on empowering women artisans and smallholder farmers while preserving Indigenous culture.	Wonosobo, Jawa Tengah; Jawa Timur; Flores

Table 3.2 (continued)

No	Company name	Business description	Area of impact	Products	Location
8	Ekoluxe & Love Khaos	A sister brand produces eco-friendly loungewear and bold luxury garments in a carbon-neutral, sweatshop-free studio, emphasizing sustainability and slow fashion. A sister brand produces eco-friendly loungewear and bold luxury garments in a carbon-neutral, sweatshop-free studio, emphasizing sustainability and slow fashion.	Eco-Friendly Fashion and Sustainable Production	Eco-friendly loungewear and luxury garments produced in a carbon-neutral, sweatshop-free studio, emphasizing slow fashion.	Bali
9	Faithfull the Brand	A socially responsible garments brand that are handmade by the local communities with a nostalgic nods to vintage patterns and fabrications.	Ethical Fashion and Fair Labor Practices	Socially responsible garments handmade by local communities, preserving traditional skills and promoting fair labor practices.	Bali
10	Miko Bahtera Nusantara (MYCL)	a biotechnology startup that develops mycelium (vegan) leather as an alternative to animal leather	Innovation in Sustainable Materials	Mycelium (vegan) leather as an alternative to traditional animal leather, focusing on reducing the environmental impact of leather production.	Bandung

Based on the SI of Adam concept (2016) and analyzing all media, data, and information distributed digitally, this study attempts to identify the characteristics of sustainable innovation in these SMEs, as indicated in Table 3.3.

Table 3.3 Content of "strategy" of SMEs for sustainability

Company	Data insight	Interpretation
Bhineka Rahsa Nusantara (BRN)	*"Rahsa Nusantara marks our commitment as a certified B Corporation to continue to uphold positive social and environmental impacts with standardized processes."* *"Rahsa Nusantara is part of the Global movement towards a more inclusive, fair, and regenerative economic system."* *"Rahsa Nusantara marks our commitment as a certified B Corporation to continue to uphold positive social and environmental impacts with standardized processes."* "every local has its wisdom" *"We aim to ensure that Rahsa Nusantara customers truly receive natural and healthy ingredients for themselves, their families, and the earth."* *"Let's join this movement so Indonesian families can live healthily and sustainably."* *"Rahsa Nusantara cooperates with the diversity of rahsa wealth, spirit, knowledge, expertise, wisdom, skills, and expertise to lead to a better world."*	Rahsa Nusantara is committed to standardized processes to ensure positive social and environmental impacts, indicating BRN's dedication to regulatory standards and efficient operational practices (OO) Rahsa Nusantara considers Indonesian families a potential market for the business opportunity of providing consumers with healthy products for improved well-being through the empowerment of local wisdom and women. (TO) Rahsa Nusantara realized that they need work together to achieve their mission. (SB) **Conclusion:** The market-driven approach with convenience and empowering local women (TO→OO, SB)

Table 3.3 (continued)

Company	Data insight	Interpretation
Aliet Green LTD	*"We've always been sustainably driven, ensuring our brand is grounded in improving our people's well-being while creating resilient communities, all concerning our environment."* *"Our women-owned business was born with a mission to create a better future every day... We're working side by side with our farmers to empower them with resilient regenerative organic farming practices that help them tackle the effects of climate change."*	Alert Green makes sustainable products (TO) Alert Green demonstrates a sustainable-driven approach to improving community well-being, works with local farmers to create their resilience (SB) **Conclusion:** societal building with a market-driven approach (SB➔ TO)
Instellar Indonesia	*"We are a purpose-driven company that helps businesses be more impactful, sustainable, and scalable."* *"We accelerate social innovation in organizations through consultancy, program, and investment by incorporating three main excellences: impact, communication, and operation."* *"We believe that purpose is not just a statement on our website or a marketing campaign; it is the driving force behind everything we do."* *"We build an impact ecosystem for collaboration between stakeholders."*	Empower business to work with sustainability (SB) Consulting businesses for sustainable business (TO) **Conclusion:** societal building with a market-driven approach (SB➔ TO)
Kmana	*"From cutting to sewing and hammering rivets, our bags and accessories are entirely hand-made, and all the materials are locally sourced."* *"With sustainability and integrity at the very heart of their creative and purpose-driven studio..."* *"Their value lies in their exceptional quality and finishing details, the know-how of the noble materials we use, and the skills and meticulousness of the craftswomen and men we partner with, like Kadek in the picture."*	The sourcing materials locally and hand-making its bags and accessories (TO) Working with local craftswomen ➔ SO Conclusion: Market driven by empowering women (TO➔ SO)

(continued)

Table 3.3 (continued)

Company	Data insight	Interpretation
Lusana	*"All dyes used are non-toxic and in full compliance with OEKO-TEX® Standard 100."* *"LUSANA's mainstay fabrics are natural and biodegradable."* *"LUSANA approaches sustainability as a never-ending process that constantly challenges the team to innovate better practices while moving beyond compliance."* *"LUSANA offers tactile collections of pieces that celebrate the female form, worn with comfort and ease."* *"The brand develops and forms partnerships with local change makers to create impact felt on-the-ground at the community level."*	Lusana's use of non-toxic dyes compliant with OEKO-TEX® Standard 100 underscores (OO) Sustainable fashion for women (TO) Partnership with local change-maker (SB) Conclusion: Market-driven with updates to the standard of sustainable fashion and collaboration (TO➔OO, SI)
Gree Energy	*"We make biogas projects financially viable by unlocking the full potential of carbon crediting programs, renewable energy, and green finance."* *"Gree Energy empowers the food industry in emerging countries to cut methane emissions, treat wastewater, and create opportunities for clean energy in underserved rural communities."* *"Our innovative business model has received international support and recognition from international development agencies and government agencies."* *"Gree Energy works with food processors from tapioca, palm oil, fruit processing to animal farming."* *"Our 2030 Targets: We Think Big - More than 1 TWh/year of clean energy."*	Gree Energy demonstrates an innovative strategy for regulatory compliance while optimizing financial efficiency (OO) Creating renewable energy (TO) Working with local food processors (SB) **Conclusion**: green energy, through getting resources from local food processors with compliance with the regulations and green finance (TO➔OO, SB)

(continued)

Table 3.3 (continued)

Company	Data insight	Interpretation
SukkhaCitta	*"In 2023, we became the first fashion brand in Asia to have our net-zero targets verified by the Science Based Target Initiative (SBTi) to keep ourselves accountable."* *"Through continuous training, we give them tools to change their lives – business skills, eco-friendly practices, quality management."* *"We exclusively use plant-based fibers and materials in our products, except for some non-vegan components like our buttons made from waste Mother-of-Pearl in Makassar and vegetarian batik wax made from sustainably sourced beeswax, pine gum, and reclaimed paraffin."* *"Our work impacts over 1500 lives, from farm to closet, increasing their income by 60%."* *"As a social enterprise, their purchases fund our grassroots projects directly."*	SukkhaCitta's achievement of net-zero targets verified by the Science Based Target Initiative (SBTi) ➔ OO Material: waste of pear, vegetarian batik from sustainable resources (TO) Increase the income of local (SB). Conclusion: Businesses from reused/recycled product form waste mother of pear in Makasar and address net-zero (TO➔SB, OO)
Ekoluxe & Love Khaos	*"NO sweatshops, NO child labor, & NO discrimination."* *"We've been 100% carbon neutral since 2020 and sweatshop free since day 1."* *"Faithfull is committed to creating thoughtfully produced, socially responsible garments that you can feel great about wearing on the inside and out."* *"we have taken measures reflecting this by introducing sustainable and certified fabrications to our collections, reducing excess fabric wastage within our production processes, using biodegradable packaging for all e-commerce orders, and implementing biodegradable garment bags throughout our supply chain this year."* *"Faithfull uses certified fabrications and chooses fibers with lower environmental impact."* *"We only work with independent, local artisans for anything we can't do in-house, such as dye our fabrics, cast our metal hardware finishings, and make our shipping supplies."*	Ekoluxe & Love Khaos' commitment to ethical practices, sustainable fashion (OO) Empower local artisan Sustainable fashion as core business value (TO) Conclusion: Sustainable fashion (TO-- > OO, SB)

(continued)

Table 3.3 (continued)

Company	Data insight	Interpretation
Faithfull the Brand	*"Faithfull is proud to be a certified B Corporation, based on our commitment to our community, our conscious business practices, and our pledge to be a force for good."* *"Our sustainability strategy is rooted in our vision and mission to grow sustainable materials for a better life."* *"Choosing to recycle agri-forestry waste rather than disposing of it in landfills is a practical and environmentally responsible choice."* *"From day one, co-directors Sarah-Jane Abrahams and Helle Them-Enger have collaborated with Bali's best artisans to create their hand-made garments."* *"Working with more than 1000 artisans across Bali, we are proud supporters of local manufacturing and preserving the artisanal skills of Indonesia."*	Faithfull's certification as a B Corporation demonstrates its commitment to regulatory compliance and ethical business practices. This certification validates its efforts to be a force for good in its community while ensuring operational efficiency (OO). Local handmade or large-scale factories (TO) Empower local artisan (SB) Conclusion: TO➔ SB, OO
Miko Bahtera Nusantara (MYCL)	*"Our raw materials didn't contain harmful chemicals, often found in conventional leather, such as Chromium Hexavalent in the tannery processing."* *"Our sustainability strategy is rooted in our vision and mission to grow sustainable materials for a better life."* *"We've developed a greenhouse gas calculator to monitor emissions in real-time across our materials and products."* *"Collaborating closely with brands and designers, we've produced responsible products that consume 70% less water, reduce global warming potential by 68%, and use 17% less energy than traditional bovine leather."* *"Our sustainability strategy is rooted in our vision and mission to grow sustainable materials for a better life."*	MYCL's avoidance of harmful chemicals in their raw materials and developing a greenhouse gas calculator illustrate their commitment to regulation ➔ OP Business as substitution of enimal based fashion, less water fashion (TO) Conclution: TO➔ OO

(continued)

Sustainable Innovation Characteristic in Indonesia B Corp SMEs

Strategy

Insights from Indonesian SMEs emphasize the importance of a robust business proposition that empowers local capacity to tap into potential markets. These SMEs identify market opportunities in various sustainability issues, such as promoting a healthy lifestyle. For instance, Bhineka Rahsa Nusantara (BRN) sees the growing trend toward a healthy lifestyle as a business opportunity and sources its primary raw materials from local Indonesian products. This commitment to sustainability is also evident in several fashion brands holding a B Corp certificate. For example, Kmana emphasizes preserving cultural heritage to promote sustainable fashion, while Gree Energy, operating in the energy sector, identifies business opportunities in renewable energy by utilizing agricultural waste. Similarly, MYCL has positioned itself as an alternative to the traditional leather industry by offering sustainable products.

A notable trend among these companies is the procurement of raw materials from local or vulnerable communities. Many SMEs prioritize delivering value over controlling the entire business supply chain. This approach is reflected in their commitment to working closely with primary stakeholders, such as farmers and local artisans, evidenced by statements like "working side by side with our farmers" and "collaborating with local artisans." These businesses aim to establish a new social system, particularly for their primary stakeholders in the supply chain. Their mission is to "improve the community's resilience" and "educate consumers for a new lifestyle," indicating a strong focus on social impact.

With such a critical mission, these businesses adhere to regulations and ethics not merely as requirements but as reflections of their values and commitment. Compliance is integrated into their strategy to streamline operations and foster a more ethical business environment.

Process

This study identifies innovation in business processes at three levels among small and medium-sized enterprises (SMEs), see Table 3.4.

- Firstly, there is a focus on **inbound logistics**, where SMEs emphasize sourcing raw materials from sustainable and certified sources. Enterprises such as BRN, Aliet Green, Lusana, SukkhaCitta, Ekoluxe, and Love Khaos demonstrate a strong commitment to sustainability by empowering local communities, including farmers, artisans, weavers, and creatives.

Table 3.4 Content of "Process" of SMEs for sustainability

Company	Data insight	Interpretation
Bhineka Rahsa Nusantara (BRN)	*"The innovations we passionately develop aim to bring forth beneficial concoctions."* *"However, Rahsa Nusantara faces challenges, such as relatively high product prices, difficulty in offline access, and a lack of recreational products."* *"Our continuously crafted services are not just for the well-being of humans but also for the Earth, a dream we nurture with gratitude."* *"Rahsa Nusantara started as a beverage business by a husband and wife in Bandung."* *"A lifelong learner so it means we are a learner, and we will never stop learning and this makes us want it improve ourself, want to grow ourself, want to develop ourself and that is actually how we really connected with the team."* *"Rahsa Nusantara cooperates with the diversity of rahsa wealth, spirit, knowledge, expertise, wisdom, skills, and expertise to lead to a better world."*	Rahsa Nusantara develops beneficial concoctions but faces challenges like high prices, limited offline access, and lack of recreational products. (OO) The company focuses on well-being for humans and Earth. (OT) Collaborates with diverse expertise to promote food sustainability and create a better world. (SB)
Aliet Green	*"Founded in 2009, Aliet Green pioneered organic coconut sugar made from regenerative organic crops, long before others learned of our practices."* *""Our women-owned business was born with a mission to create a better future every day."* *"Collaboratively working with smallholder farmers, research institutes, educational organizations, and individuals, together we'll achieve sustainable and ecological food systems, superior labor conditions and ethical working procedures that empower rural communities to create their own successes whilst respecting nature and all of their habitats."*	Regenerative Organic Crops practice (OO) Sustainable food systems and ethical labour practices by empowering rural communities and women (SB)
Instellar Indonesia	*"Instellar designs tailored programs together with partners by utilizing our expertise in incubation, acceleration, strategic planning, and investment readiness."* *"Our approach, like a symphony, is driven by data and understanding."* *"Our investment thesis is prosperous people and planet."* *"We want to build a world where businesses are run with social responsibility and driven by purpose and impact."* *"We build an impact ecosystem for collaboration between stakeholders."*	Tailored Programs for Partners (OO) Purpose-driven business committed to social responsibility to reshape future paradigm (OT) Impact Ecosystem (SB)

Table 3.4 (continued)

Company	Data insight	Interpretation
Kmana	*"Guided by the motto 'less is more', we embrace clean and subtle detailing."* *"We also believe in responsible profits and are committed to full price transparency by offering a detailed breakdown."* *"Soon, we will announce that we have been formally certified as a B Corp. We couldn't be prouder."* *"We work with small workshops and cooperatives of artisans, weavers, and creatives..."* *"We partner with small family-run tanneries (like the Ken run by Pak Eko, in the picture) in East Java."*	Minimalistic and refined design production (OO) Committed to ethical profitability and transparency (OT) Partnering with local artisans and promoting community empowerment (SB)
Lusana	*"LUSANA approaches sustainability as a never-ending process that constantly challenges the team to innovate better practices while moving beyond compliance."* *"In 2022, LUSANA began supporting the transition of community farmers toward regenerative practices through subsidizing a seaweed-based organic fertilizer."* *"LUSANA is a human-centered brand that values each relationship —whether within our team, among our vendors, or outwardly with our growing loyal customer base."* *"The brand develops and forms partnerships with local change-makers to create impact felt on the ground at the community level."* *"LUSANA operates a centralized supply chain that is based in Indonesia. The brand works directly with Tier 1, Tier 2, and sometimes Tier 3 suppliers."*	LUSANA innovating sustainable practices beyond compliance (OO) Prioritizing human relationships and values (OT) Collaboration for community-level impact (SB)
Gree Energy	*"Gree Energy provides proven and dependable biogas solutions that deliver reliability, high performance and safety."* *"With decades of combined experience in biogas, we deliver unrivaled and profitable biogas solutions."* *"Gree Energy decarbonizes agriculture and food supply-chains at scale."* *"Our solutions are compliant with world-class industry standards and frameworks including Verified Carbon Standard (VCS), IFC's Environmental and Social Performance Standards (IFC PS), Sustainable Development Goals (SDGs), and IRIS+."* *"Gree Energy bridges the gap for most food processors in emerging countries, making biogas technologies accessible and affordable."*	Provides reliable, high-performance biogas solutions (OO) Decarbonizes food supply chains sustainably with compliance (OT) Bridges gap for food processors with biogas technology (SB)

(continued)

Table 3.4 (continued)

Company	Data insight	Interpretation
SukkhaCitta	*"We started growing our own cotton and dyes regeneratively in 2020, working directly with smallholder farmers, and plan to regenerate 1000 Hectares of land by 2030."* *"Through continuous training, we give them tools to change their lives – business skills, eco-friendly practices, quality management."* *"We work tirelessly to raise awareness and create demand for these products."* *"We're proud to have shipped our products to over 30 countries, sustaining our village-level work."* *"We work directly with smallholder cotton and indigo farmers to reclaim their land—we plan to regenerate 1000 Hectares of land by 2030."* *"We're a social enterprise dedicated to ending the exploitation of women in villages across Indonesia while caring for our environment."*	Regenerates land and uses natural cycles for sustainable fashion (OO) Empowers farmers with skills and promotes eco-friendly practices (OT) Ships products globally, supporting community-level regeneration efforts (SB)
Ekoluxe & Love Khaos	*"We believe you can always do both. By that, we mean fashion and function should play well together; you can be comfortable and well-dressed."* *"We believe in a fashion industry that values people, the environment, creativity, and profit equally."* *"We ask that you compost and/or recycle our packaging when finished."*	Extensive efforts for maximum environmental sustainability (OO) Promotes fashion that blend comfort, style, and ethical values (OT) Encourages recycling and composting of packaging materials (SB)
Faithfull the Brand	*"We are putting in the time to improve and do better and will share this with our community as we understand the importance of transparency and measurability."* *"We are committed to becoming the most responsible version of ourselves."* *"Faithfull the Brand is extremely proud to be a certified B Corporation, based on our commitment to our community, conscious business practices, and pledge to be a force for good."* *"We work closely with local manufacturers and artisans in Bali, Indonesia, and beyond, and implement ethical and responsible practices across the business."* *"Weekly sustainability meetings have been implemented with our team for progress checkpoints to ensure we are keeping on track."* *"We pay all our 100+ employees, in Indonesia and Australia, above national minimum wage, along with additional benefits*	Investing in technology to reduce fabric wastage, enhancing sustainability (OO) Committed to sustainability with regular team meetings and achieving B Corporation certification (OT) Passionate about supporting local communities and implementing ethical practices in collaboration with manufacturers and artisans (SB)

Table 3.4 (continued)

Company	Data insight	Interpretation
Miko Bahtera Nusantara (MYCL)	*"We've just launched a new fundraising round seeking $four million USD to propel us to the next level."* *"Recently, we closed our pre-Series A funding round with investors from fashion, agriculture, and life sciences sectors joining our cap table."* *"Since our founding in 2015, our values in conducting business have always been people, planet, and prosperity."* *"We actively collaborate with Indonesian farmers for raw material sourcing and partner with local designers and artists to create customized products."* *"Our efforts have garnered recognition, including social impact awards such as MIT Solve Seed and the DBS Foundation Award."*	Seeking and successful funding (OO) Founded on people, planet, and prosperity values, targeting leather industry pollution (OT) Collaborating with Indonesian farmers and local artists, winning social impact awards (SB) Conclusion: Funding for sustainable impact OO → OT, SB

- Secondly, **operations and product creation** are centered on meeting regulatory standards while prioritizing environmental management and sustainability. Some SMEs have pursued B Corp certification and have emphasized promoting human rights within their organizational practices. For instance, Faithfull the Brand underscores its commitment to ethical practices with policies such as "NO sweatshops, NO child labor, & NO discrimination," alongside ensuring fairness in employee compensation.
- Lastly, there is a focus on **outbound logistics**, where SMEs position themselves as educators, promoting better lifestyles and environmentally friendly products to consumers. Many of these enterprises aim to offer fair pricing, as exemplified by Kmana's commitment to providing reasonable prices for their consumers. Additionally, these SMEs promote inclusion and sustainability through their marketing campaigns, particularly in the fashion industry. Brands like Kmana, Lusana, SukkhaCitta, and Faithfull the Brand utilize social media to communicate their values, positioning themselves as fair companies that prioritize ethical and sustainable practices. The content of learning of SMEs is shown in Table 3.5.

Learning

It is evident from their social media activity and public statements that maintaining a learning culture is central to these SMEs' mission, as they regularly update and refine their sustainable practices. Their commitment to fostering a learning culture is demonstrated in several key ways:

- **Firstly**, they build strong relationships within their supply chain, as shown through their communications and photos shared on social media. This engage-

Table 3.5 Content of "Learning" of SMEs for sustainability

Company	Data Insight	Interpretation
Bhineka Rahsa Nusantara (BRN)	*"This togetherness represents all the hopes, sweat, and laughter of all people who process & work in Rahsa Nusantara."* *"Rahsa Nusantara's contribution strengthens efforts in sustainable women's empowerment through various programs and business operations."*	Embodies collective aspirations, dedication, and camaraderie among its workforce, reflecting their shared commitment and efforts (OT) Sustainable women's empowerment through integrated programs and business practices (SB)
Aliet Green LTD	*"We set out to champion more small family holders and female and disabled agroforestry farmers and improve our Indigenous communities' well-being."*	Engage with small family, women, and disabilities (OT)
Instellar Indonesia	*"We support UNDP in building business and impact measurement capacity for startups and SMEs, focusing on the blue sector."* *"Instellar Program partners with visionary organizations to develop impactful and tailor-made programs."* *"Instellar designs tailored programs with partners, utilizing our expertise in incubation, acceleration, strategic planning, and investment readiness."*	Capacity building for Startups and SMEs in the Blue Sector (OO) Engage with partners to support Social Enterprises (OT) Assist Sustainable Business Model development (SB)
Kmana	*"All the workshops and cooperatives we work with have been visited and verified..."* *"Positive luxury is what they call it. Creative and conscious, Kmana makes a bag for every person and every occasion."*	Kmana rigorously verifies all partner workshops and cooperatives, emphasizing creative and conscientious luxury craftsmanship. (OT)
Lusana	*"LUSANA is a human-centered brand that values each relationship — whether within our team, among our vendors, or outwardly with our growing loyal customer base."* *"The brand develops and forms partnerships with local change-makers to create impact felt on the ground at the community level."*	Prioritizes human-centered relationships and community-level partnerships for impact (OT)

Table 3.5 (continued)

Company	Data Insight	Interpretation
Gree Energy	*"Gree Energy is the only B Corp certified biogas solution provider in Asia and meets the highest standards of verified social and environmental performance."* *"Gree Energy partners with food processing facilities to unlock a sustainable and cost-effective future."* *"Gree Energy delivers financially viable and accessible biogas solutions leveraging over 50 years of experience in biogas technology, carbon credits, and renewable energy asset development."*	Highest social and environmental B-Corp Standards (OO) Partnerships for sustainable future with processors (OT) Biogas Solutions with extensive expertise (SB)
SukkhaCitta	*"For years, we looked to Nature to find better ways to do things. Through countless hours of research, ratios, and recipes, we finally found a way to harness its natural power to create vibrant colors."* *"Through continuous training, we give them tools to change their lives – business skills, eco-friendly practices, quality management."* *"We're not just another fashion brand or business; we're a change model."*	Asia's first fashion brand with verified net-zero targets, uses natural dyes (OO) Utilizes indigenous knowledge for plant-based colors, empowers communities (OT) Combats women's exploitation, champions environmental care, and advocates for social change (SB)
Ekoluxe & Love Khaos	*"In short, YES, we have spent hours, upon weeks, upon months, on end (literally) to ensure we have taken every precaution to be as environmentally sustainable as humanly possible."* *"As 1% of the planet's members, we donate money to vetted programs and organizations that help restore the environment."* *"As the message spreads, we slowly start to change minds and open eyes about the fashion industry and how we can all do a little better."*	Supports environmental restoration through vetted program and B Corporation certification (OT)
Faithfull the Brand	*"As Faithfull continues to grow, our ongoing goal is to see these small, independent factories, and the people working within them, prosper even further."*	Faithfull emphasizes the prosperity of small, independent factories and ensures fair wages and benefits for all employees, promoting well-being and sustainable growth (OT)

(continued)

Table 3.5 (continued)

Company	Data Insight	Interpretation
Miko Bahtera Nusantara (MYCL)	*"Throughout our journey, we've been supported by invaluable partners, including local and global research facilities like Future Service Laboratory, MIT, Zurich University, and SU TV Singapore."* *"We're proud to have secured pilot partnerships with global brands like Decathlon, Under Armour, and Camper to launch capsule collections next year."* *"We're proud to have secured pilot partnerships with global brands like Decathlon, Under Armour, and Camper to launch capsule collections next year."* *"Recently, we closed our pre-Series A funding round with investors from fashion, agriculture, and life sciences sectors joining our cap table."*	Supported by global research, piloting with major brands (OO) with sustainable values (people, planet, and prosperity) (OT) Secured global brand partnerships and closed funding round to innovate further (SB)

ment allows them to understand the practical issues at the forefront of their business processes. Their mission to improve the well-being of their supply chain partners, such as farmers and local artisans, brings them closer to local wisdom and fosters an understanding of how to work together to create shared values.

- **Secondly**, they stay updated with industry-related sustainability information. For example, Ekoluxe and Love Khaos focus on understanding trends within the fashion industry. Additionally, participating in B Corp certification helps them stay informed about evolving sustainability trends relevant to their work.
- **Thirdly**, these organizations have developed a robust learning system that includes regular workshops, training sessions, and meetings to discuss environmental and social issues. Brands like Kmana, SukkhaCitta, and Ekoluxe & Love Khaos exemplify this approach. They also have mechanisms for continuous improvement, such as evaluation systems that help them refine their practices over time.

Furthermore, their strategy includes collaborating with new partners to stay abreast of emerging trends. This approach allows them to gain new insights from different perspectives. For instance, MYCL collaborates with institutions like Future Service Laboratory, MIT, Zurich University, and SU TV Singapore, which positively impacts their product development and market strategy. The content of linkage of SMEs for sustainability is shown in Table 3.6.

Table 3.6 Content of "Linkage" of SMEs for sustainability

Company	Data insight	Interpretation
Bhineka Rahsa Nusantara (BRN)	*"Rahsa Nusantara cooperates with the diversity of rahsa wealth, spirit, knowledge, expertise, wisdom, skills, and expertise to lead to a better world."* *"Rahsa Nusantara is a local start-up from Indonesia with a vision, mission, and desire to be a sustainable enterprise from the start."*	Leverages a diverse range of resources to foster sustainability and embodying its origins as a local Indonesian startup committed to sustainable practices from the outset (OT)
Aliet Green LTD	*"Collaboratively working with smallholder farmers, research institutes, educational organizations, and individuals, together we'll achieve sustainable and ecological food systems, superior labor conditions and ethical working procedures that empower rural communities to create their own successes whilst respecting nature and all of their habitats."*	Collaborate with locals to enhance sustainable food system (OT) Empower rural communities (SB)
Instellar Indonesia	*"Instellar Program partners with visionary organizations to develop impactful and tailor-made programs."* *""Mentors meet with social enterprises monthly to provide advice and feedback.* *"Instellar emphasizes collaborative efforts with stakeholders to achieve sustainable development goals and foster economic empowerment."* *"We support UNDP in building business and impact measurement capacity for startups and SMEs, focusing on the blue sector."* *"The program addressed enterprise needs and challenges."* *"Collaborations: We highly embraced the collaboration between teams, not limited to the cohort, but also with external parties within Instellar network and IKEA co-workers."*	Partnering with impact organiszation for programs (OO) Collaborates with partners for building business and impact measurement (OT) Address enterprise needs within Instellar's network (SB)
Kmana	*"We don't want to integrate them as laborers vertically; we want to work with them as partners."* *"From a spirit of travel, Kmana was born. A brand of sustainable bags founded in Bali in 2018, where craftsmanship and creating opportunities for local communities are the foundations of its responsible business in the fashion industry."* *"We are deeply committed to slow fashion principles; sustainability is part of our DNA."*	Collaborate with mutual respect and sustainable relationships with local communities (OT) Engage with stakeholders based on the principles of slow fashion to minimize garments toxin (SB)

(continued)

Table 3.6 (continued)

Company	Data insight	Interpretation
Lusana	*"The brand develops and forms partnerships with local change-makers to create impact felt on the ground at the community level."* *"The brand develops and forms partnerships with local change-makers to create impact felt on the ground at the community level."*	Prioritizes natural, biodegradable fabrics and community-level partnerships for impact (OT) Prioritizes natural, biodegradable fabrics and community-level partnerships for impact. (SB)
Gree Energy	*"Gree Energy's business model has received funding and technical assistance from prominent intergovernmental organizations such as the UN program Seed Capital Assistance Facility, the International Renewable Energy Agency, and European development aid agencies."* *"Gree Energy extends gratitude to our dedicated team, strategic partners, and supporters worldwide whose support has been instrumental in our journey of excellence in sustainability."* *"Gree Energy empowers food processors in emerging countries to deploy biogas solutions that treat industrial wastewater and cut methane emissions."*	Supported by partners (UN, IRENA, and European Agencies) (OO) Acknowledges team, partners, and supporters globally (OT) Empowers processors with biogas for wastewater and emissions (SB)
SukkhaCitta	*"We work directly with smallholder cotton and indigo farmers to reclaim their land—we plan to regenerate 1000 Hectares of land by 2030."* *"We exclusively use plant-based fibers and materials in our products, except for some non-vegan components like our buttons made from waste Mother-of-Pearl in Makassar and vegetarian batik wax made from sustainably sourced beeswax, pine gum, and reclaimed paraffin."*	Direct engagement with smallholder farmers, global reach in shipping (OT) Empowers stakeholders with training, emphasizes sustainable materials (SB)
Ekoluxe & Love Khaos	*"We work with local artisans in Denpasar (where our studio is) who use a natural process to hand dye our fabrics in small batches."* *"We are passionate about being part of the slow fashion movement and proud to be part of the solution instead of part of the problem."* *"With your conscious efforts, we can MASSIVELY minimize this carbon footprint."*	Collaborates with local artisans for natural fabric dyeing (OO)Embraces slow fashion, aiming to be part of the environmental solution (OT) Urges collective efforts to reduce carbon footprint (SB) significantly

Table 3.6 (continued)

Company	Data insight	Interpretation
Faithfull the Brand	*"Our primary focus is our local community in Bali, Indonesia; however, we also provide support to international organizations based on global circumstances and the needs of the communities."* *"We have always been focused on implementing a transparent chain of production – working with local manufacturers and artisans in Bali, Indonesia, and beyond."*	Prioritizing local Bali community, with international support (OT) Emphasizing transparency and continuous improvement efforts (SB)
Miko Bahtera Nusantara (MYCL)	*"Collaboration with sustainable experts from national research institutes, World Resources Institute, Solar Impulse Foundation, B Lab, and Fashion for Good has guided us in planning and implementing our sustainability initiatives."* *"Our efforts have garnered recognition, including social impact awards such as MIT Solve Seed and the DBS Foundation Award."* *"Our sustainability strategy is rooted in our vision and mission to grow sustainable materials for a better life."* *"We actively collaborate with Indonesian farmers for raw material sourcing and partner with local designers and artists to create customized products."*	Guided by sustainable experts, recognized with awards (OT) Sustainability strategy: sourcing, collaborating, creating customized products (SB)

Linkage

Two key characteristics of linkages implemented by these small and medium-sized enterprises (SMEs) are:

- **Firstly**, collaborations are integrated into the business value chain, particularly in the upstream supply chain. This involves partnerships with farmers, artisans, and local communities to support and enhance the sustainability of their business operations. All SMEs sampled in this study engage in such collaborations with the aim of strengthening their operational lines and maximizing their impact on these stakeholders.
- **Secondly**, these collaborations are focused on enhancing sustainability capabilities, both environmentally and socially. They are not limited to domestic institutions but also extend to international partners. For example, Instellar Indonesia collaborates with the UNDP, while MYCL partners with the World Resources Institute and the Solar Impulse Foundation to amplify their impact. Similarly, Aliet Green works with educational and research institutions to bolster their sustainability efforts. The content of "Innovative Organization" of SMEs for sustainability is shown in Table 3.7.

Table 3.7 Content of "Innovative Organisation" of SMEs for sustainability

Company	Data Insight	Interpretation
Bhineka Rahsa Nusantara (BRN)	"Since 2016, Rahsa Nusantara has been dedicated to an ethical and conscientious approach, emphasizing the use of natural, local ingredients, environmental mindfulness, and empowerment." "The innovations we passionately develop aim to bring forth beneficial concoctions." "Rahsa Nusantara is a local start-up from Indonesia with a vision, mission, and desire to be a sustainable enterprise from the start." "Rahsa Nusantara provides a platform that offers natural and healthy consumer packaged goods, education for healthier and sustainable living, digital communities, and as a model of a more circular business model for production and consumption."	Focuses on ethical practices with local ingredients and innovative solutions (OO)Integrates Natural Benefits into its ethical production process (OT) Promoting Circular Business Model (SB)
Aliet Green LTD	*"We always step forward to innovate and provide better leadership, creativity, competence, climate, and efficiency solutions."*	Innovate to achieve the best solution for the upcoming challenges (OT)
Instellar Indonesia	*"Our approach, like a symphony, is driven by data and understanding."* *"Instellar designs tailored programs with partners, utilizing our expertise in incubation, acceleration, strategic planning, and investment readiness."* *"We believe innovative ideas can solve issues like pollution, waste, and poverty through business approaches."* *"We are a funding partner who provides innovative financing instruments that match Zebra's stripes."* *"we aim to build a better Indonesia by supporting these businesses in various ways."*	Utilize data-driven strategies (OO) Funds innovative solutions (OT) Integrating profitability with sustainability (SB)

Table 3.7 (continued)

Company	Data Insight	Interpretation
Kmana	*"We created Kmana to become a bridge between Southeast Asian craftsmanship traditions and the contemporary scene..."* *"Kmana is a family-run, purpose-driven creative studio based in Bali and Singapore."* *"We are now exploring how to close the loop and move towards a circular model..."* *"We wanted to redefine the concept of 'Made in Indonesia', aspiring to product quality and giving it the value it deserves."*	Craftmanship with contemporary practices (OT) Circular economy to redefine "Made in Indonesia" (SB)
Lusana	*"In 2022, LUSANA began supporting the transition of community farmers toward regenerative practices through subsidizing a seaweed-based organic fertilizer."* *"LUSANA approaches sustainability as a never-ending process that constantly challenges the team to innovate better practices while moving beyond compliance."* *"LUSANA approaches sustainability as a never-ending process that constantly challenges the team to innovate better practices while moving beyond compliance."* *"As a fashion brand that values natural fabrications, our ready-to-wear items originate from farming or forestry-based sources."*	Supports farmers with regenerative practices, innovating sustainably (OO)Sustains innovation beyond compliance in sustainability practices (OT)Innovates sustainably with natural fabrications, fostering environmental responsibility (SB)
Gree Energy	*"Gree Energy provides customized solutions for biogas projects, offering a comprehensive set of services from feasibility studies to EPC to green financing."* *where environmental responsibility is paramount."* *"Gree Energy fights climate change by scaling profitable biogas solutions that transform food industry waste into valuable resources for people and the planet."*	Customizes biogas solutions with full-service offerings (OO) Commits to sustainability and pushing boundaries (OT) Scales biogas solutions from waste to resources (SB)

(continued)

Table 3.7 (continued)

Company	Data Insight	Interpretation
SukkhaCitta	*"We started growing our own cotton and dyes regeneratively in 2020, working directly with smallholder farmers, and plan to regenerate 1000 Hectares of land by 2030."* *"We're not just another fashion brand or business; we're a change model."* *"We're a social enterprise dedicated to ending the exploitation of women in villages across Indonesia while caring for our environment."* *"We're not just another fashion brand or business; we're a change model."*	Regenerative farming, plant-based materials, sustainable components (OO) SukkhaCitta raises awareness, drives demand, and models change (OT) Addressing exploitation and environmental care (SB)
Ekoluxe & Love Khaos	*"We make the most eco-conscious product possible..."* *"Designing and manufacturing an environmentally friendly product is something we're passionate about."* *"We are proud to be part of the slow fashion movement and proud to be part of the solution instead of part of the problem."* *"Our designs are striking, investment pieces created with longevity in mind because we encourage you to curate a wardrobe of select, well-made pieces you love instead of piles of clothing you will only wear once."* *"Slow fashion is the movement of designing, creating, and buying garments for quality and longevity."*	Focuses on creating the most environmentally friendly product possible (OO) Passionate about designing and manufacturing environmentally friendly products (OT) Proudly embraces slow fashion, advocating for quality over quantity in clothing consumption (SB)
Faithfull the Brand	*"We are very excited to invest in technology that will reduce excess fabric wastage even further, which will be introduced to the business by the end of 2021."* *"It is challenging for anyone in the fashion industry to consider themselves completely sustainable, and we have the utmost respect for those who have achieved this title."*	Investing in technology to reduce fabric wastage (OO) Acknowledging sustainability challenges in fashion, proud B Corp certification (SB)

Table 3.7 (continued)

Company	Data Insight	Interpretation
Miko Bahtera Nusantara (MYCL)	*"Recently, we closed our pre-Series A funding round with investors from fashion, agriculture, and life sciences sectors joining our cap table."* *"Our team comprises individuals from diverse backgrounds, united by a shared goal of creating a better future for generations to com"* *"Our sustainability strategy is rooted in our vision and mission to grow sustainable materials for a better life."* *"Our team comprises individuals from diverse backgrounds, united by a shared goal of creating a better future for future generations."* *"Choosing to recycle agri-forestry waste rather than disposing of it in landfills is a practical and environmentally responsible choice."*	Diverse investors and future-focused team (OO) Sustainable materials for future generations (OT) Recycling agri-forestry waste for sustainability (SB) Conclusion: Integration of diverse perspective and innovative strategies to advance sustainability goals OO ➔ SB, OT

Innovative Organisation

Small and medium enterprises (SMEs) leverage innovation to achieve sustainability in three distinct ways:

- **First, Innovation in Business Processes:** This involves innovating various business processes, from sourcing raw materials to delivering products to the market. All the companies discussed in this study exhibit this behavior. For instance, Sukkhacita is developing organic cotton and engaging in regenerative farming with local farmers, aiming to reforest 1000 hectares of land. Conversely, Kmana is focused on designing its production process to create a circular business model.
- **Second, Product Innovation:** Many SMEs are making a significant environmental impact by developing alternative products to those currently available on the market. Gree Energy and MYCL are notable examples of such SMEs. Gree Energy operates in the renewable energy sector, while MYCL is transforming the fashion industry with its biological approach to product creation.
- **Third, Social Innovation:** Most SMEs in this study have sought to empower stakeholders within their supply chains, including farmers, artisans, and individuals with specific needs, such as tailors. These brands are working to enable stakeholders to adopt new practices, particularly within the supply chain. This approach not only aims to make products more environmentally friendly but also integrates social innovation to enhance overall impact. For example, LUSANA supports community farmers' transition to regenerative practices by subsidizing seaweed-based organic fertilizer. Aliet Green assists farmers in producing organ-

ically sourced palm sugar. Even Instellar Indonesia has incorporated this initiative into its business model by partnering with the UNDP.

- **Finally, Innovation in Business Models.** The aforementioned innovations have created unique opportunities for SMEs to refine their business models. Each change in the innovation process indirectly influences the character of their business model, positioning these SMEs for future growth.

Discussion and Conclusion

Indonesian B-Corp-certified SMEs are emerging as transformative entities within their sectors, demonstrating a strong commitment to integrating sustainable practices and developing innovative business models. Their strategic focus on sustainability, particularly within their supply chains, underscores their dedication to creating positive social and environmental impacts.

Strengths and Innovations

1. **Commitment to Sustainability and Values**

   The primary strength of B-Corp-certified SMEs lies in their unwavering commitment to sustainability. These organizations adeptly identify and exploit market opportunities linked to sustainability. For instance, Bhineka Rahsa Nusantara (BRN) leverages the growing consumer trend toward healthy lifestyles by sourcing raw materials locally, thus supporting Indonesian agricultural products. Similarly, Gree Energy capitalizes on the renewable energy sector by utilizing agricultural waste, while MYCL offers an eco-friendly alternative to traditional leather products. These examples illustrate the SMEs' strategic alignment with sustainability trends across diverse sectors.

   A notable trend among these SMEs is their emphasis on sourcing raw materials from local or vulnerable communities. This practice is evident in their partnerships with farmers, artisans, and other local stakeholders, aiming to bolster their supply chains and create shared value. This approach not only reinforces the supply chain but also contributes to the economic resilience of local communities. The commitment to these relationships is reflected in statements such as "working side by side with our farmers" and "collaborating with local artisans," highlighting the SMEs' dedication to improving community well-being and fostering new social systems.

   Moreover, these SMEs view regulatory compliance and ethical practices as integral to their operational strategies rather than mere obligations. For example, Faithfull the Brand's policy of "NO sweatshops, NO child labor, & NO discrimination" underscores their commitment to ethical labor practices and fairness in compensation, aligning with their broader mission and values.

2. **Innovation Across Business Processes**

   The study reveals a nuanced approach to innovation among B-Corp-certified SMEs, categorized into three integral levels: inbound logistics, operations and product creation, and outbound logistics.

   At the level of inbound logistics, B-Corp-certified SMEs exhibit a pronounced commitment to sustainable sourcing practices. This commitment is exemplified by companies such as BRN and Aliet Green, which prioritize the procurement of raw materials from certified, sustainable sources. These SMEs actively engage with local communities, fortifying their supply chains and contributing to broader sustainability objectives. By sourcing materials locally and working closely with communities, these organizations enhance their operational sustainability while supporting regional economies.

   In the realm of operations and product creation, these SMEs focus on aligning their practices with regulatory standards and integrating environmental management into their processes. This approach is evident in their pursuit of B Corp certification, which serves as formal recognition of their adherence to rigorous sustainability criteria. For instance, Kmana and SukkhaCitta demonstrate their commitment through the development of circular business models and regenerative farming practices, respectively. Kmana's efforts in designing a circular business model reflect a proactive stance towards minimizing waste and maximizing resource efficiency, while SukkhaCitta's engagement in regenerative farming highlights a focus on environmental stewardship and the restoration of ecological balance.

   The approach to outbound logistics among B-Corp-certified SMEs is characterized by a robust educational component aimed at promoting sustainable lifestyles and environmentally friendly products. These SMEs leverage marketing strategies to advocate for ethical consumption and inclusivity. Brands such as Kmana and Faithfull the Brand utilize social media platforms to articulate their values, including fair pricing and ethical labor practices. Through these channels, they not only communicate their commitment to sustainability but also influence consumer behavior by positioning themselves as proponents of both social and environmental responsibility.

3. **Learning and Adaptation**

   A commitment to fostering a learning culture is central to the strategic approach of B-Corp-certified SMEs, manifesting in several key areas that support ongoing development and innovation.

   One critical aspect of these SMEs' strategies involves actively building and maintaining relationships within their supply chains. Through continuous communication and shared experiences, these organizations address practical challenges and gain a deeper understanding of local knowledge and practices. This engagement not only helps in resolving immediate operational issues but also facilitates the creation of shared value among stakeholders. By integrating local wisdom into their business processes, SMEs enhance their ability to operate sustainably and effectively within their communities.

Another crucial element of their learning culture is staying informed about industry trends and evolving sustainability practices. Participation in B Corp certification exemplifies this commitment, as it requires ongoing engagement with up-to-date sustainability standards and practices. Companies such as Ekoluxe and Love Khaos demonstrate this approach through their active efforts to remain current with industry developments. They regularly partake in workshops and training sessions, which serve to reinforce their knowledge and adapt their practices in alignment with the latest trends and innovations in sustainability.

Strategic collaborations with leading institutions further enhance the learning culture of these SMEs. Partnerships with prestigious organizations such as MIT and the World Resources Institute provide valuable insights and support continuous improvement in business models. For instance, MYCL's collaborations with Future Service Laboratory, MIT, Zurich University, and SU TV Singapore exemplify how these alliances contribute to product development and market strategies. These partnerships enable SMEs to leverage external expertise and resources, driving innovation and refining their approaches to sustainability.

Challenges and Barriers

Despite their strengths, these SMEs face significant challenges. Market penetration remains a critical hurdle, as Indonesian consumers often perceive sustainable products as expensive. With a substantial portion of the population falling into the middle to low-income brackets, these businesses encounter difficulties in reaching a broader audience. This challenge is compounded by the fact that many of these SMEs operate in traditional sectors, such as food and fashion, where sustainability may not yet be a priority for consumers.

Additionally, transforming stakeholder behavior presents another major challenge. Many stakeholders, including farmers and artisans, have limited access to education and resources, which impedes their ability to adopt new, sustainable practices. Overcoming cultural barriers and persuading stakeholders to embrace change requires significant investment in education and capacity-building.

Business Model and Learning Culture

The innovation in business models among B-Corp-certified SMEs is another critical strength. These SMEs often prioritize service and shared value, which helps integrate stakeholders with varying levels of skill and education into the new economic landscape. Their commitment to continuous learning and improvement is evident in their daily practices, where learning is not confined to top-level management but is integrated throughout the organization. This approach facilitates the refinement of business models and enhances their impact.

Implications for Future Economy

Despite the challenges, the efforts of these SMEs to promote a sustainable mindset are likely to positively influence the future economy. By setting an example of integrating sustainability into business practices, they can drive broader industry changes and shape consumer attitudes toward sustainable products. Their innovative approaches to sustainability and commitment to continuous learning position them as leaders in driving positive economic and social transformations.

Strategic Framework for Enhancing Sustainable Innovation in B-Corp-Certified SMEs

To enhance their impact and efficacy, B-Corp-certified SMEs could benefit from adopting a more explicit strategic framework. The following recommendations are proposed to build on their existing strengths and address observed challenges:

1. **Strengthening Inbound Logistics through Strategic Sourcing Partnerships**
   To optimize inbound logistics, SMEs should deepen their strategic sourcing partnerships. Building on practices like those of BRN and Aliet Green, SMEs should formalize these partnerships through long-term agreements and joint initiatives. Developing supplier codes of conduct aligned with B Corp standards can enhance consistency in sustainability practices and reinforce supply chain resilience. Implementing blockchain technology for traceability could increase transparency and accountability in sourcing practices.
2. **Advancing Operations and Product Creation through Circular Economy Models**
   SMEs should strategically integrate circular economy principles into their business models. Companies like Kmana and SukkhaCitta provide exemplary models with their focus on circular production and regenerative farming. SMEs should adopt circular design principles that prioritize product longevity, recyclability, and resource efficiency. Establishing cross-industry collaborations to share best practices and innovative technologies can drive significant advancements. Investing in research and development to explore new sustainable materials and production processes will further enhance their industry position.
3. **Enhancing Outbound Logistics with Consumer Education Initiatives**
   Outbound logistics strategies should include targeted consumer education initiatives. As demonstrated by brands such as Kmana and Faithfull the Brand, effective use of social media and other communication channels can significantly influence consumer behavior. SMEs should develop comprehensive educational campaigns highlighting the environmental and social benefits of their products. Partnering with educational institutions and NGOs to co-create content can amplify their reach and impact. Implementing feedback mechanisms to assess consumer understanding and attitudes toward sustainability will allow SMEs to tailor their messaging more effectively.

4. **Institutionalizing a Learning Culture through Formalized Programs**
   SMEs should establish formalized programs for continuous education and professional development. Creating structured programs that include regular workshops, training sessions, and knowledge-sharing forums will enhance their strategic approach. Partnering with academic institutions and industry experts can provide valuable insights and foster innovation. Developing an internal knowledge management system to capture and disseminate learnings across the organization will ensure alignment with sustainability goals.
5. **Leveraging Strategic Collaborations for Innovation and Market Expansion**
   Strategic collaborations with leading institutions and industry partners should be formalized to drive innovation and market expansion. SMEs should seek partnerships that offer complementary expertise and resources. Engaging with international research organizations and industry leaders can provide access to cutting-edge technologies and global market insights. Establishing joint ventures or strategic alliances can facilitate entry into new markets and enhance the company's competitive edge. Participating in industry consortia focused on sustainability can position SMEs as thought leaders and advocates for broader systemic change.

By implementing these strategic recommendations, B-Corp-certified SMEs can enhance their operational efficiency, drive innovation, and amplify their impact on sustainability. These strategies offer a clear pathway to addressing existing challenges and seizing new opportunities for growth and leadership in sustainable business practices.

In conclusion, Indonesian B-Corp-certified SMEs are making significant strides in transforming their industries through sustainable innovation. Their commitment to mission-driven values and innovative business models, coupled with their efforts to address market and stakeholder challenges, highlights their role as change agents in the future economy.

References

Adams, R., Jeanrenaud, S., Bessant, J., Denyer, D., & Overy, P. (2016, 4). Sustainability-oriented innovation: A systematic review. *International Journal of Management Reviews, 18*(2), 180–205.

Anggarini, L. T. (2020). The innovation in business model for sustainability in biotechnology startups in Indonesia. *International Journal Enginering Technology*, 130–137.

Asad, M., Rizwan, A., Shah, M., & Munir, A. (2018, 11). Impact of innovation practices on sustainable performance SMEs. *Herald National Academy of Managerial Staff of Culture and Arts, 3*, 537–546.

Assink, M. (2006, 4). Inhibitors of disruptive innovation capability: A conceptual model. *European Journal of Innovation Management, 9*(2), 215–233.

B Lab. (2020a, 5). *B Impact Assessment Structure*. Retrieved from https://kb.bimpactassessment.net/support/solutions/articles/43000574682-b-impact-assessment-structure

B Lab. (2020b, 5). *Impact areas: Governance, workers, community, environment and customers*.

Bocken, N., Short, S., Rana, P., & Evans, S. (2014, 2). A literature and practice review to develop sustainable business model archetypes. *Journal of Cleaner Production, 65*, 42–56.

Ceschin, F. (2013, 4). Critical factors for implementing and diffusing sustainable product-Service systems: Insights from innovation studies and companies' experiences. *Journal of Cleaner Production, 45*, 74–88.

Famiola, M. (2020). SMEs' social and environmental initiatives in Indonesia: an institutional and resource-based analysis. *Social Responsibility Journal, 16*(1), 15–27.
Gunawan, A., Bloemer, J., Van Riel, A., Essers, C., Bastian, L., Al Dajani, H., et al. (2022, 7). Institutional barriers and facilitators of sustainability for Indonesian Batik SMEs: A policy agenda. *Sustainability, 14*(14), 8772.
Hallstedt, S., Thompson, A., & Lindahl, P. (2013, 7). Key elements for implementing a strategic sustainability perspective in the product innovation process. *Journal of Cleaner Production, 51*, 277–288.
Haruuma, I. (2023, 1 24). *Sanksi Membakar Sampah Sembarangan*. Retrieved 03 23, 24, from https://nasional.kompas.com/read/2023/01/25/01000001/sanksi-membakar-sampah-sembarangan?page=all
Muangmee, C., Dacko-Pikiewicz, Z., Meekaewkunchorn, N., Kassakorn, N., & Khalid, B. (2021, 4 13). Green entrepreneurial orientation and green innovation in small and medium-sized enterprises (Smes). *Social Sciences, 10*(4).
Permatasari, D., Yulianto, A., Mohammed, N., & Shafie, N. (2023, 11). Factors influencing business sustainability of SMEs in Indonesia post Covid-19 pandemic. *Proceedings of the international conference in technology, humanities and management (ICTHM 2023), 12–13 June, 2023, Istanbul, 131*, 167–178.
Setyaningsih, S., Widjojo, R., & Kelle, P. (2024, 12). *Challenges and opportunities in sustainability reporting: A focus on small and medium enterprises (SMEs)*. Retrieved from https://www.tandfonline.com/doi/abs/10.1080/23311975.2023.2298215
Silvestre, B., & Ţîrcă, D. (2019, 1). Innovations for sustainable development: Moving toward a sustainable future. *Journal of Cleaner Production, 208*, 325–332.
Teece, D. (2007, 12). Explicating dynamic capabilities: The nature and microfoundations of (sustainable) enterprise performance. *Strategic Management Journal, 28*(13), 1319–1350.
Yuliani & Soetjipto, B. (2019). Why should SMEs in Indonesia have to apply "Green Practices"? *Advances in Economics, Business and Management Research, 101*.

Giorgio Gatica Singgih, MBA, is a dedicated professional with expertise in social enterprises, CSR, and ESG. He holds a Master of Business Administration from SBM ITB. Giorgio has a strong background in project management, leadership, and organizational development. He currently leads Kasih Balik, an organization focused on youth empowerment, education, and community building through collaborative projects. His experience includes roles as a People Advisory Services Consultant at EY and multiple research assistantships with organizations such as EBSI, UNDP, and AKATIGA. Giorgio is committed to making a positive social impact through his work.

Melia Famiola, STP, MT, PhD, is a lecturer at the School of Business and Management, Bandung Institute of Technology. She has been asked to serve as the PRME Coordinator at SBM ITB and is also involved in the management of PRME Asean. Currently, Melia is dedicating her research to the central theme of transforming our society towards sustainability, particularly in Asian countries.

Chapter 4
Innovative Marketing Strategy for Green Products in Search of Better Customer Acquisition

Dina Dellyana and Leo Aldianto

Abstract In this chapter, we examine the critical role that sustainability plays in modern marketing strategies, driven by consumer awareness and regulatory requirements. The study examines how businesses are integrating eco-friendly practices to enhance their brand reputations and ensure long-term viability, including reducing carbon emissions. Companies that use green marketing improve their market position by strengthening their relationships with environmentally sensitive consumers and adhering to rules. In order to build consumer loyalty and trust, the research also highlights the significance of environmental stewardship, social responsibility, and transparency in marketing.

According to the findings, successful green marketing techniques have a big influence on consumer behaviour, which boosts advocacy and brand loyalty. It is advised that sustainability initiatives be regularly highlighted through transparent and lucid communication through a variety of channels, that consumers be educated about the environmental advantages of eco-friendly products, and that interactive experiences that fully immerse customers in the brand's sustainable objectives be developed. The chapter also recommends that companies concentrate on long-term involvement by offering post-purchase assistance and fostering a community around common interest.

Keywords Sustainability · Marketing · Consumer awareness · Regulatory requirements

D. Dellyana (✉) · L. Aldianto
School of Business and Management, Institut Teknologi Bandung, Bandung, Indonesia
e-mail: dina.dellyana@sbm-itb.ac.id; leo.aldianto@sbm-itb.ac.id

A. H. Verkuil et al. (eds.), *Core Values and Decision-Making for Sustainable Business*, Sustainable Business Development,
https://doi.org/10.1007/978-3-031-78361-6_4

Introduction

Businesses are increasingly realizing the significance of sustainability due to growing consumer demands, regulatory requirements and the necessity to address negative impacts. In regard to the environment, companies can contribute to greenhouse gas emissions, resource depletion, pollution, and harm ecosystems. In the social sphere, poor labor practices, unsafe working conditions, and inadequate community engagement can negatively impact employees and the community. In terms of economics, unsustainable business models and unfair trade practices can erode local economies and cause financial instability. Through sustainable practices, businesses can meet regulatory requirements and consumer expectations while also enhancing their long-term success and resilience.

Companies are embracing methods like reducing carbon emissions utilizing eco materials and prioritizing social responsibility. This shift goes beyond compliance; it aims at gaining an advantage enhancing brand reputation and ensuring long term viability. Consequently, marketing strategies need to adapt to these values by highlighting transparency, social accountability and environmental stewardship. Green marketing, which emphasizes the promotion of products and services based on their environmental benefits, differs significantly from traditional marketing approaches due to its focus on sustainability and environmental stewardship. Unlike conventional marketing, which primarily aims to drive sales and increase market share, green marketing seeks to align a company's offerings with the growing consumer demand for environmentally responsible products. This approach requires companies to not only highlight the eco-friendly attributes of their products but also to ensure that their overall business practices are sustainable and transparent. Green marketing entails educating consumers about a company's sustainability initiatives integrating these practices into all business functions and fostering lasting consumer loyalty by aligning with their beliefs (Winston et al., 2019). Through efforts businesses can forge connections, with environmentally conscious consumers strengthen their market position and contribute positively to global sustainability objectives. According to a study by Rahahleh et al. (2020), consumers' green buying intentions are significantly influenced by their perception of a company's environmental commitment and transparency. The importance of authenticity in green marketing, however, is highlighted by Mehraj and Qureshi (2020) in order to avoid the pitfalls of greenwashing and ensure consumer trust in the process.

As the focus of green marketing research has evolved, there has been a discernible movement in research patterns from broad green concepts to more specialized fields like industrial and environmental technologies (Jung & Kim, 2023). Many factors are driving the growth of green marketing research. Research in this field has also been influenced by shifts in consumer behavior and the growing demand for eco-friendly products. Academic curiosity is piqued by the necessity to comprehend how customers view and react to green marketing initiatives in order to design tactics that work (Jung & Kim, 2023).

Climate change, environmental degradation, and economic inequality are pressing issues that call for immediate action to prevent social unrest, conflict, economic disruptions, and the possible collapse of the socio-political-economic systems that have been built over centuries, according to growing scientific evidence and first-hand observations (Yunus & Desai, 2017). Researchers are looking into how marketing might help address global environmental and social difficulties as a result of people's growing awareness of these issues. As a result, there is an increasing amount of study on green marketing strategies that support social responsibility and environmental stewardship (Jung & Kim, 2023).

In addition, numerous local, regional, and federal governments have made a commitment to sustainability by establishing goals, rules, and legislation to promote socially and environmentally conscious actions and business ventures (Saravanan et al., 2018; Wijen et al., 2012). Corporate marketers are acting in tandem with the growing public awareness over social and environmental issues. Many different industries have started initiatives to involve stakeholders and follow social and environmental standards; these include 3 M, Dow Chemicals, Kimberly-Clark, Microsoft, Novelis, Procter & Gamble, Unilever, and Walmart (Apte & Sheth, 2016; Bhattacharya & Polman, 2016; Villena & Gioia, 2020; Winston et al., 2019).

Sustainability was recently highlighted by international investment management company BlackRock in its yearly letter to CEOs of businesses in its portfolio (Krämer, 2020). These patterns suggest that there will probably be a rise in the number of stakeholder actions related to corporate compliance with sustainability goals, due diligence, and ethical business practices. These initiatives still represent an acknowledgment of the need to address social and environmental challenges, even in the face of concerns about "greenwashing," "blue washing," and "green development goals washing"—where businesses downplay negative effects while marketing their positive contributions to sustainability (Parguel et al., 2011).

Green Customer

A "green consumer" is someone who believes that each person has a personal responsibility for the environment. Government is not the sole entity accountable; the business community and society at large also bear some of the blame. Businesses are often blamed for contributing to environmental issues through practices such as pollution, resource depletion, and unsustainable production methods. Similarly, society at large, including consumers, is blamed for unsustainable consumption patterns and waste generation. By recognizing this shared responsibility, green consumers advocate for more sustainable practices across all sectors, encouraging businesses to adopt eco-friendly operations and individuals to make environmentally conscious choices. Their thought process is quite deep; they consider current events as well as apply them to their actions. Five groups were identified based on research done on American consumers by Roper Stach Worldwide: TrueBlue Greens, Greenback Greens, Sprouts, Grousers, and Basic Browns. According to

Table 4.1 Consumers categorization based on their level of environmental consciousness

Green area	Characteristics
True Blue Greens	They possess strong environmentalist values. They initiate positive change and strive to teach environmentalist values.
Greenback Greens	They are not politically active environmentalist consumers. However, they perform environmentally friendly consumption more frequently than an average consumer.
Sprouts	They theoretically believe in green consumption; however, they practically do not materialize these opinions.
Grousers	They do not have any information on environmental issues. They do not believe to create differences through their behaviours.
Basic Browns	They do not change their behaviours due to environmental or social issues or do not feel sorry for these issues.

Source: Adapted from Ginsberg-P and Bloom (2004)

Ginsberg and Bloom's research, there exists a subset of consumers who exhibit more environmentalist behavior than others in society. These consumers can be categorized into five groups based on their level of environmental consciousness (Table 4.1).

Green customers are consumers who are environmentally conscious and prefer to purchase products and services that are eco-friendly and green. They consider the environmental impact of a product throughout its lifecycle, from production to disposal, and are willing to pay a premium for green offerings. Green customers value products made from recycled or renewable materials, with minimal packaging, energy-efficient, and having a low carbon footprint. Their purchase decisions are driven by a desire to reduce their environmental footprint and support companies that prioritize sustainability practices. Understanding the attitudes, behaviors, and motivations of green customers is crucial for businesses to develop effective green marketing strategies, product designs, and communication campaigns that resonate with this growing segment of conscious consumers (Srisathan et al., 2023; Herman & Herman, 2014).

Green Marketing Versus the Conventional Marketing

Since green marketing is more comprehensive and incorporates social, economic, and environmental factors into business plans, it requires a different approach than traditional marketing. Through the promotion of goods and methods that are socially and ecologically conscious, green marketing seeks to produce long-term value in contrast to conventional marketing, which is primarily concerned with increasing sales and short-term profits. This strategy necessitates a mental change away from only meeting the wants of consumers and toward taking into account the welfare of society and the environment. Green marketing entails building consumer trust, promoting responsible consumption, and being open and honest about a business's sustainability initiatives. However, some consequences may be faced by the company when starting to implement green practice. Usually, it requires upfront investments in sustainable technologies, processes, and materials. The company may have to invest in renewable energy sources, energy-efficient equipment, sustainable raw materials, and eco-friendly packaging. Supply chain management can be made more difficult by sourcing environmentally friendly goods and making sure suppliers follow regulations. Cost increases and disruptions to the supply chain may result from this. Also, navigating the complex landscape of environmental regulations and standards can be challenging and time-consuming. Non-compliance can result in legal penalties and damage to the company's reputation.

In order to guarantee that sustainability is integrated into every phase of product development and delivery, cooperation throughout the supply chain is also required. Company entering green practice need to build strong relationships with suppliers and work collaboratively to ensure adherence to sustainability standards, including diversify the supplier base to reduce dependency on any single source and enhance supply chain resilience. To answer the funding, companies may start to seek external funding sources such as green loans, grants, or investment from sustainability-focused investors. Green practices can be phased in gradually over time to spread out the costs. In green marketing, companies should provide verifiable evidence and third-party certifications to ensure authenticity and transparency.

Businesses can enhance brand loyalty and contribute to a more environmentally conscious future by developing deeper and more genuine relationships with consumers that emphasize sustainability through several approaches to green marketing. Marketing plays an important role in society by educating consumers about products and services that can improve their quality of life. However, marketing can only continue to exist if its communication tactics successfully inform, educate, and steer consumers toward green solutions. To create awareness of green products and assist customers in making informed decisions, eco-friendly products need to be properly communicated. Promoting ecological awareness, advertising, sales promotion, public relations, and personal selling using a variety of promotion methods is recommended" (Banyte et al., 2015). This incorporates the presumption that marketing initiatives also seek to change wasteful and hazardous consuming practices, such as the post-consumption disposal and recycling of products. Positive incentives, new product and process innovations, deterrents against specific consumer practices through pricing or other means, and macromarketing activities that constructively engage numerous stakeholders are all necessary to transform, and occasionally even reduce, consumption patterns (Little et al., 2019; Peattie & Peattie, 2009; Scott et al., 2014).

This calls for a shift in perspective, cutting-edge resources, and organizational modifications. Green marketing necessitates conscientious and progressive business tactics that direct the market toward environmentally friendly products and services. It is feasible for businesses to pursue economically sustainable practices while upholding social and environmental standards, despite the considerable challenges involved. Many businesses have shown they can handle seemingly incompatible goals. Take total quality management (TQM), for example, where they effectively reduced costs while increasing quality (Curkovic et al., 2000; Schmidheiny, 1992). While the path to green development may appear challenging and drawn out, we think that businesses can successfully create a green marketing strategy by focusing on four essential areas: rethinking goods and services, encouraging responsible consumption, modifying the marketing mix, and restructuring the supply chain to ensure sustainability at every stage of production and distribution.

The entire marketing mix, from product and packaging to positioning and promotion, must be completely redesigned for green marketing. Marketers need to assess how the production, content, packaging, labeling, advertising, distribution, usage, and disposal of their products will affect sustainability. Beyond this evaluation, there exist tactical prospects to react favorably and win over environmentally conscientious customers, advocacy organizations, the media, and merchants. These changes could involve the creation of new products as well as potential acquisitions. Strong sustainability marketing strategies have already been put into place by retailers such as H&M, Ikea, Nordstrom, Sephora, Target, and Walmart, as well as brands like Levi's, Patagonia, and Reformation (Widlitz, 2020).

Repackaging, relabeling, reformulating, and repositioning are all part of redefining the marketing mix. There is a growing concern regarding packaging that harms the environment and excessive packaging. These days, a lot of marketers use recycled or biodegradable packaging. Retailers are beginning to put products on display that have packaging that has been relabeled with additional information on sustainability. Additionally, manufacturers are repurposing goods to improve energy efficiency or get rid of potentially dangerous components like phosphates in detergents. Products that are positioned as environmentally friendly are more visible in the congested advertising area and are more likely to be recognized by consumers who care about the environment.

Redefining sales strategies and incentive plans is also necessary. Providing clients with information on environmental issues can have a big effect. Giving clients access to critical sustainability data on suppliers gives you a competitive edge in business-to-business settings. Selling companies will face more pressure as organizations prioritize responsible sourcing and expect sustainability from their suppliers. Retailers are already giving preference to suppliers whose goods improve their reputation as advocates for sustainability. Proactive leadership and ethical selling have many more advantages than disadvantages.

Green Product

A green product is one that has been engineered to reduce its adverse effects on the environment over the course of its whole life cycle. It is also referred to as an ecologically friendly or eco-friendly product. It usually utilizes minimal packaging, conserves energy and natural resources during manufacturing and use, and produces little to no pollution or harmful waste because it is made mostly of recycled, renewable, or biodegradable components. Green products take into account environmental considerations from the sourcing of raw materials to the manufacturing processes, distribution, consumption, and end-of-life disposal or recycling. This helps them reduce their carbon footprint, conserve ecosystems, and promote sustainability. Energy-efficient appliances, organic food, environmentally friendly cleaning supplies, and goods manufactured from recycled or sustainable materials are a few examples. The fundamental idea behind green products is to maximize benefits to customers while reducing harm to the environment (Sheth & Parvatiyar, 2021; Winston et al., 2019).

Companies need to think about the effects of their products at every stage of their lifespan, from manufacture to disposal, in addition to innovating new products and services. They ought to concentrate on reducing threats to the environment, making the best use of available resources, and encouraging customers to adopt "eco-efficiency" and "eco-sufficiency" practices (Heikkurinen et al., 2019). A green economy should make "reconsumption"—the using and reusing of products over several cycles or generations—its industrial objective. A major duty for marketers is to create items that are long-lasting and encourage repurposing while also informing customers of their advantages. Due to their greater robustness and reusability, items produced of high-density polymers, ceramics, and rare metals may have a less negative environmental effect than those made of steel, aluminum, or other metals (Andrady, 2015; Vatan & Yilmaz, 2020). Businesses who thrive in this field and convince customers of the advantages will have an advantage over rivals as public awareness of sustainability rises. Lifecycle thinking and management must therefore be integrated into the circular economy (Lacy & Rutqvist, 2015).

Utilizing the lifespan entails taking responsibility for it. State of the art marketing does not end with the sale. Hence, marketing managers are usually very deeply connected to their markets, strategies and long-term objectives. To manage this duty, comprehensive conceptual and operational strategies are required, particularly as governments throughout the world start to control product disposal at the end of its lifecycle and packaging. Competitive benefits can come from reconsumption-promoting strategies. It has been observed that eco-friendly products can fetch up to a 25% premium, reflecting consumers' willingness to pay more for sustainable options (Lin & Huang, 2012). For instance, customers frequently pay more for environmentally friendly travel experiences and food that is cultivated organically (Pulido-Fernández & López-Sánchez, 2016).

Green Place

The conventional notion of the 'Place' component in the marketing mix has generally been linked to the administration of product distribution channels or physical service locations. According to Pretty et al. (2005), a significant portion of the environmental effect of tangible consumer goods is related to the energy and resource consumption used in these distribution channels as well as the consumer transit to and from retail locations. The development of environmentally friendly distribution channels has received a lot of attention in marketing scholarship and practice research. Although this affects pricing and costs associated with distribution, customers are generally not the main focus. There is an exception in the food sector, where consumers are becoming more interested in food miles and preferring to consume more 'local' food. The function that merchants play in forming relationships with customers and influencing their behavior through sustainable marketing techniques is a crucial topic of research concerning product distribution (Jones et al., 2011).

Green Promotion

Green Advertisement

A green marketing strategy involves promoting green products, services, or brand images. Consumers have become increasingly aware of environmental issues in recent years and are seeking eco-friendly alternatives. The goal of green advertising is to attract environmentally conscious consumers and demonstrate a company's commitment to environmental responsibility. Some important aspects of green advertising include:

- Product attributes: Making use of environmentally friendly materials, reducing emissions, or optimizing energy use (Dangelico & Vocalelli, 2017).
- Production processes: Emphasizing sustainable manufacturing methods, renewable energy usage, or waste reduction efforts (Kumar, 2016).
- Corporate image: Showing commitment to environmental issues and sustainability (Leonidou et al., 2011).
- Eco-labels and certifications: Using certified or eco-labelled products to confirm their environmental credentials (Taufique et al., 2019).

Green advertising can be found in a variety of places, such as digital platforms like social media and websites, print and broadcast media, and product packaging. However, businesses need to exercise caution to avoid "greenwashing," which is the practice of presenting fictitious or overstated environmental statements that harm the reputation of a brand (de Jong et al., 2018).

A company's brand image can be enhanced, consumer loyalty can rise, and it may even gain a competitive edge in the market with the help of efficient green advertising (Kumar & Christodoulopoulou, 2014). To preserve customer trust and stay out of trouble with the law, companies must make sure that their environmental claims are supported by evidence and consistent with their real operations.

Green Personal Selling

Green personal selling is a sales technique that combines the conventional personal selling method with ecologically responsible ideas. This strategy focuses on promoting goods and services that are environmentally sustainable through direct interactions between sales professionals and potential clients. In addition to making sales, the objective is to inform clients about the product's positive environmental effects and the business's dedication to sustainability. Green personal selling works very well in consumer goods, energy, and automobile industries, where products have a big environmental impact. In addition to standard marketing methods, sales staff must be trained in environmental concerns, sustainable practices, and the particular eco-friendly aspects of the goods or services they are pushing.

Some important elements of green personal selling are:

- Product Knowledge: According to Sharma and Iyer (2012), sales people need to be extremely knowledgeable about the advantages and environmental features of the goods they are selling.
- Environmental Education: Salespeople should inform clients about environmental issues and how their products address them.
- Ethical Selling Practices: To prevent greenwashing and overstated environmental claims, place an emphasis on honesty and transparency.
- Relationship Building: Concentrating on establishing enduring bonds with clients who care about the environment (Guo et al., 2017).
- Customization: Adapting the sales presentation to each customer's or market segment's unique environmental concerns (Sharma & Iyer, 2012).

Green Sales Promotion

The term "green sales promotion" describes the application of eco-friendly marketing techniques to stimulate the purchase of eco-friendly goods and services. This tactic is a component of a larger green marketing plan that attempts to match advertising with objectives related to environmental sustainability. Offering samples, discounts, and hosting eco-friendly events are just a few strategies that can be used in conjunction with other green sales campaigns to draw in and involve customers who care about the environment.

Important components in promoting green sales include:

- Discounts and Coupons: To encourage purchases, offer discounts or coupons for eco-friendly products (Rahman et al., 2017).
- Samples: Giving away free samples of environmentally friendly items so that customers can test them before deciding whether or not to buy them
- Putting together eco-friendly events that highlight green products and raise awareness of environmental issues (Rejeki et al., 2015).

- Publicity: Making use of media attention to draw attention to the advantages that products have for the environment and the company's dedication to sustainability (Hossain & Rahman, 2018).
- Educational campaigns: disseminating knowledge about how products affect the environment and promoting environmentally friendly customer behavior (Tiwari et al., 2011).

The Green Public Relation

Green Public Relations (Green PR) is a specialized form of public relations that focuses on communicating an organization's environmental initiatives, sustainability practices, and corporate social responsibility efforts related to environmental issues. Enhancing an organization's reputation, fostering stakeholder confidence, and advancing ecologically friendly practices are the objectives of green public relations.Some important facets of green PR are:

- Spreading the word about sustainability initiatives: showcasing a company's initiatives to lessen its influence on the environment, such cutting carbon emissions or starting recycling programs (Ottman, 2011).
- Engaging a variety of stakeholders, such as clients, staff members, investors, and local communities, in order to raise awareness of environmental issues and get input on sustainability initiatives is known as stakeholder engagement (Rawlins, 2006).
- Crisis management: To preserve public trust, environmental challenges or crises must be handled responsibly and transparently (Seeger, 1997).
- Media relations: cultivating connections with media channels to disseminate uplifting narratives on the company's environmental endeavours
- Campaigns for education: raising public knowledge of environmental challenges and encouraging sustainable lifestyles (Tiwari et al., 2011).
- Preventing greenwashing: To prevent charges of deceptive or fraudulent advertising, make sure that all environmental statements are true and supported by evidence (Chen & Chang, 2012).

Green Sponsorship

Green sponsorship is the practice of providing cash or in-kind support for environmental projects, events, or initiatives with the goal of strengthening corporate social responsibility and sustainability. This strategy is advantageous to the sponsoring organization as well as the environment.

Important Green Sponsorship Aspects

- Contributing money to environmental organizations or projects is known as financial support (Korindo Foundation, 2021).

- Contributions in kind: Providing items or services that aid in sustainability projects; for example, supplying eco-friendly goods or services for events
- Event sponsorship: Endorsing occasions that emphasize sustainability and environmental consciousness, as green conferences or eco-friendly festivals
- Corporate Social Responsibility (CSR): Showing a dedication to environmental stewardship by including green sponsorship into larger CSR efforts

Green Direct Marketing

The practice of green direct marketing consists of using direct marketing techniques to promote environmentally friendly products and services while minimizing their environmental impact. Email, direct mail, social media, and SMS are all direct communication channels that can be used to reach environmentally conscious consumers. Among the key strategies are using eco-friendly materials for physical mailings, such as recycled paper and non-toxic inks, as well as offering digital alternatives. To avoid accusations of greenwashing, green direct marketing emphasizes transparency and authenticity, which ensures that claims about the environment are substantiated. Rahmat et al. (2023) assert that companies can engage with green consumers and reinforce their environmental responsibility through direct marketing efforts that include sustainable practices.

Green Point of Purchase Communication

Green point of purchase (POP) communication is the term for educational and marketing materials and campaigns that emphasize sustainability and are put on display in retail establishments where customers make purchases. The goal of this kind of communication is to encourage consumers to make more sustainable decisions at the point of sale and to highlight eco-friendly items.

Important facets of green POP communication consist of:

- Emphasizing the benefits to the environment: POP materials highlight the environmentally friendly features of items, like their energy efficiency, recyclable packaging, or lower carbon footprint (Ummar et al., 2023).
- Clear and clear messaging: Concise, understandable language is used in effective green POP communication to quickly communicate sustainability information to consumers (Ummar et al., 2023).
- Consumer education: Green POP displays frequently offer details on sustainability concerns and the ways in which the product responds to them (Tunley Environmental, 2023).
- Visual cues: To indicate environmental friendliness, green hues, images of nature, and eco-labels are frequently utilized (Tunley Environmental, 2023).

By using green POP marketing effectively, firms may stand out from the competition, win over ecologically sensitive customers, and possibly even see a rise in sales of sustainable goods. To prevent being accused of greenwashing, businesses must make sure their claims are true and substantiated.

Green Social Media Campaign

Social media platforms increasingly play a significant role in peoples' daily lives. Social media communication expands not only from one individual to another but also from one individual to hundreds of people and eventually to millions of people. A green social media campaign makes use of social media channels to encourage eco-friendly behaviors, sustainability, and environmental awareness. These campaigns share interesting and educational content about environmental challenges, sustainable products, and conservation initiatives in an effort to change consumer attitudes and behaviors toward green consumerism. Clear messaging, captivating imagery, and calls to action that inspire followers to lead more sustainable lives are essential components of successful green social media initiatives (Ummar et al., 2023). However, the cost of online activity in energy terms is indeed considerable and can have a significant impact on the sustainability and "green" credentials of companies. A substantial amount of energy is required for online activities, including data storage, streaming, and cloud computing. This energy consumption contributes to the carbon footprint of companies, potentially undermining their efforts to present themselves as environmentally responsible. Businesses should invest in renewable energy sources to power their data centers to reduce their reliance on non-renewable energy and manage the negative effects of social media activities on energy consumption and carbon footprint. The use of energy-efficient technologies and practices, such as optimizing server utilization and improving cooling systems, can further reduce energy consumption. Additionally, companies should regularly monitor and report their energy use and carbon emissions to ensure transparency and accountability. By adopting these measures, businesses can align their social media operations with their sustainability goals, enhancing their green credentials and demonstrating a genuine commitment to environmental responsibility (International Energy Agency, 2020).

Green Packaging

Green packaging involves creating and using packaging solutions that have an impact on the environment throughout their lifecycle. This includes designing packaging that can be recycled, broken down naturally or produced from sources to decrease waste and preserve resources. The goal of packaging is to limit the use of renewable materials, cut down on carbon emissions during manufacturing and transportation and ensure that packaging can be easily recycled or turned into compost after it has been used. It takes into account the supply chain starting from sourcing materials to disposing of them at the end of their life cycle promoting a circular economy where materials are reused and recycled continuously. By embracing packaging practices businesses can lessen their impact comply with regulations and meet the rising consumer demand for environmentally friendly products—ultimately contributing to a more green and ethical approach, in business operations.

To make packaging greener, one business needs to adopt a holistic approach that considers the entire lifecycle of the packaging, from design and material selection to production, use, and disposal. Below are some green packaging strategies that can be carried by companies:

- Use of Sustainable Materials: To lessen packaging's negative environmental effects, use recyclable, biodegradable, or compostable materials (Martinho et al., 2015)
- Minimalist Design: Cutting down on material and packing size without sacrificing product safety
- Eco-Friendly Labeling: Informing customers by providing clear recycling guidelines and straightforward environmental information on packaging (Rokka & Uusitalo, 2008).
- Reusable packaging refers to creating packaging that customers can reuse or repurpose (Lofthouse et al., 2009).
- Innovative Materials: Making use of cutting-edge environmentally friendly materials such as plastic derived from plants or recycled ocean plastic (Geyer et al., 2017).
- Life Cycle Assessment: Carrying out thorough evaluations of packaging's environmental effects over the course of its life.
- Transparency: Giving thorough details regarding the sourcing and environmental impact of the packaging (Magnier & Schoormans, 2015).

Green packaging techniques are essential because they greatly lessen the environmental impact of products at every stage of their lifecycle, from manufacture to disposal. Recyclable, biodegradable, and compostable packaging materials reduce waste and pollution and meet customer demand for environmentally friendly products, which is on the rise. Furthermore, green packaging improves a company's reputation by showcasing its dedication to environmental responsibility, which draws in eco-aware customers and strengthens brand loyalty. Furthermore, implementing sustainable packaging techniques can boost a business's competitive edge by reducing costs through material efficiency and regulatory compliance.

In sectors like food and energy-efficient appliances, labels are important because they provide consumers with information that influences their purchasing decisions. Research has demonstrated that customer behavior can be positively impacted by sustainability labeling (Grankvist et al., 2007; Thøgersen et al., 2010). Labeling can help create a unique market position for producers and manufacturers, allowing them to highlight their commitment to sustainability and quality and eventually fostering a community of ethical consumers. Moreover, clear and credible labeling can build trust and transparency, which are crucial for maintaining long-term customer loyalty and driving repeat purchases (Testa et al., 2015).

Lastly, it can be difficult to convey complicated sustainability issues using labels. Labels must also be multilingual, adhere to an increasing number of ingredient-related regulatory requirements, and serve an expanding number of international markets. Efforts to decrease packaging often result in less area available for labelling information (Prothero et al., 1997). The efficacy of labelling is greatly increased by the use of independent certification for product claims, which is based on a comprehensive life-cycle analysis of the product's socio-environmental performance.

Green Price[1]

Green pricing, in which businesses establish prices for goods and services that are ecologically friendly, is a crucial part of green marketing tactics. With this strategy, the frequently higher costs of creating sustainable goods are intended to be balanced against consumers' willingness to pay a premium for environmentally friendly solutions. Chen and Chang (2012) state that while developing green pricing strategies, it is important to take into account variables including target market groups, manufacturing costs, and the perceived value of environmental benefits. In order to convince environmentally conscious customers to pay more, successful green pricing frequently entails being open and honest about the factors driving up costs, such as the use of sustainable resources or moral labor standards (Polonsky & Rosenberger III, 2001).

A comprehensive understanding of consumer behavior and market dynamics is necessary for the successful implementation of green pricing. According to Laroche et al. (2001), customers that care deeply about environmental issues are more willing to pay extra for green items. However, D'Souza et al. (2006) warn that even among consumers who have pro-environmental views, price sensitivity might still be a barrier. To overcome this challenge, businesses may employ strategies like highlighting the long-term financial benefits of eco-friendly solutions or offering a range of green products at different pricing ranges. Additionally, some companies use dynamic pricing methods, which adjust green premiums based on market conditions and consumer demand.

[1] Section "Green Promotion" and parts of "Green Packaging" and "Green Price" are similar to some student's paper from the ITB university. As the paper is not published, it cannot be cited.

Green Customer: Behavioural and Attitudes

Behaviour of Green Product Consumer

Green consumers demonstrate behaviour that gives environmental sustainability first priority when making decisions about what to buy and how to live. Consumers typically look for products that include eco-friendly features like energy efficiency, recyclable packaging, or organic components. These customers may actively shun goods or businesses that they believe to be detrimental to the environment, and they are frequently prepared to pay higher costs for environmentally friendly solutions. Social conventions, consumer effectiveness perception, attitudes, and environmental awareness all have an impact on their behaviour. Green consumers generally recycle, use less energy and water, and patronize firms that practice environmental responsibility. However, studies have shown a discrepancy between green views and real behaviour, with a number of obstacles, including cost, accessibility, and ease of use, influencing the conversion of pro-environmental intentions into practical actions (Joshi & Rahman, 2015; Paul et al., 2016).

Consumers' behavior in regard to green products is influenced by a number of factors. Consumers who are environmentally conscious, value personal values, and pay attention to social influences tend to prioritize environmentally friendly products. Kumar and Ghodeswar (2015) conducted a systematic literature review and found that societal norms, perceived consumer effectiveness, and environmental understanding are important variables influencing green consumer behavior. Higher environmental awareness and knowledge among consumers increases the likelihood that they will make green purchases because they recognize the beneficial effects of their decisions on the environment. Marketers can further promote green behavior by highlighting the psychological and emotional benefits of green products, such as the satisfaction of doing something good for the environment, the benefits of self-expression, and exposure to the outdoors (Liao et al., 2020).

Attitude Toward Green Brand

The term "attitude" refers to how one feels about certain environmentally friendly actions or goods, such as choosing organic food or other eco-friendly items. Consumers with strong green attitudes are more likely to prioritize sustainability in their purchasing decisions, frequently looking for goods that reduce their negative effects on the environment and support ecological preservation. According to Lee, consumers' assessment and reasoned judgment of green products has given rise to the notion of attitude toward green brands. There are some strategies to increase attitude consumers towards green brands:

- Conduct positive word-of-mouth and efficient environmental advertising (Verma et al., 2019),
- Highlight the psychological and emotional advantages of green products, such as the satisfying feeling that comes from doing good for the environment, the benefits of self-expression, and exposure to the outdoors (Liao et al., 2020).
- Take the environment into account when designing products (Borchardt et al., 2011).
- Providing clear, accessible information via a variety of platforms, including websites, social media, and in-store displays, about the benefits of buying green products and the effects their purchases have on the environment (Heimlich & Ardoin, 2008).
- Increasing the perceived value of environmentally friendly products by highlighting their long-term cost savings, high quality, and health advantages can have a beneficial impact on consumer attitudes (Chen & Chang, 2012)
- Let consumer's peers and influential personalities use and support green products (Joshi & Rahman, 2015).
- Fostering trust through openness on a company's sustainability policies and environmental impact. A comprehensive list of sourcing, production methods, and certifications can provide reassurance to customers about the validity of green promises (Ganapathi & Mahesh, 2013).
- Providing discounts, loyalty points, or other benefits to customers who buy eco-friendly items can encourage them to adopt a more favourable opinion of them (Dehghanan & Bakhshandeh, 2014).
- Focus on mid- to high-income levels of customers since they have a more favorable perception of green products and are more likely to be able to afford green items online (Srisathan et al., 2023).

Green Marketing Towards Purchase Intention and Decision

Due to growing customer demand for eco-friendly goods and services and increased environmental consciousness, green marketing has grown in significance. Businesses must comprehend how green marketing affects consumers' purchasing intentions and decisions in order to match their marketing tactics to the tastes and expectations of their target market. Businesses can improve their brand perception, gain the trust of consumers, and eventually increase sales of their environmentally friendly products by putting into practice efficient green marketing strategies. Businesses may adjust their marketing strategies and product development to match the changing demands of environmentally conscious consumers by having a thorough understanding of the elements that impact green buying behavior, such as environmental concerns, awareness of green brands, and perceived product quality. Increased client loyalty, a competitive edge, and long-term business viability can result from this understanding (Manongko & Kambey, 2018).

It is thought that a product-focused green marketing strategy can better integrate environmental concerns into all facets of business operations, from strategy development and planning to production and client distribution. It is thought that by paying attention to the green marketing approach, customer purchase decisions might be influenced. Customers make selections about what to buy based on their perception of a product, which gives them confidence that what they are doing is right. The degree of the customer's confidence in his selection of a product is reflected in his confidence in his purchase decision.

Green Purchase Intention

The term "green purchase intention" refers to consumers' willingness and likelihood to buy environmentally friendly products or services. This is often driven by their concern for environmental issues and sustainability. Purchasing decisions are influenced by an individual's propensity to consider the environmental impact of their purchases. There are many factors that influence green purchase intentions, including environmental knowledge, attitudes towards green products, perceived consumer effectiveness, social norms, and product-specific attributes, such as perceived quality and value. Despite a gap between intentions and actual behavior due to factors like price, availability, and convenience, research has indicated that consumers with higher green purchase intentions are more likely to engage in eco-friendly consumption patterns (Chen & Chang, 2012; Joshi & Rahman, 2015).

Some strategies to increase green purchase intention are:

- Enhancing the perception of value consumers receive from green products (Chen & Chang, 2012).
- Reducing perceived risks associated with green products, such as performance uncertainty or higher costs, can increase purchase intentions (Chen & Chang, 2012).
- Increasing consumer trust in green brands and products (Chen & Chang, 2012).
- Illustrating the transformation of products into environmentally friendly versions (Zhang et al., 2023).
- Informing consumers about the product's origin and production process can positively affect green purchasing decisions (Zhang et al., 2023).
- Making green brands look cool can mediate the relationship between product transformation and purchase intention (Zhang et al., 2023).
- Take demographic factors into consideration when tailoring green purchase strategies (Wang & Wong, 2019).
- Increase consumers' sense of green identity, as it can influence their decision to buy eco-friendly goods (Barbarossa et al., 2017)
- Use marketing techniques that align with collectivist principles to promote green products (Al Zubaidi, 2020).

Innovative Green Marketing and Branding Strategy

In summary, some innovative strategy to increase the intention of consumer to buy green products can be answered using method by Stengel (2011):

1. *Must eliciting joy.* Brands can utilize innovative tactics that establish gratifying emotional connections and unforgettable experiences related to sustainability to generate enthusiasm and excitement among consumers for environmentally friendly products. One strategy is to make choosing eco-friendly feel enjoyable and interesting by incorporating amusing, creative, and interactive features into the product design or packaging. For instance, use eye-catching hues, humorous illustrations, or interactive components to inform consumers about the environmental advantages of the product. Incorporating customers into sustainability projects, like tree-planting events or beach clean-ups, may also help brands cultivate a sense of community and shared purpose. This creates a sense of collective impact and joy in helping a greater cause. Customers can also feel proud and happy when the brand celebrates milestones and tells motivating tales of how its environmentally friendly products are changing the world. Brands have the power to evoke positive feelings in consumers, making them feel happy, fulfilled, and rewarded for making the eco-friendly decision.
2. *Enabling connection.* Brands that sell eco-friendly products should use innovative tactics that promote a feeling of community, transparency, and shared values in order to establish a stronger relationship with consumers. Using content marketing and storytelling to communicate the brand's sustainability journey and emphasize the good change they hope to bring about and the environmental impact of their products is one successful strategy. Behind-the-scenes tours of environmentally friendly production facilities, staff profiles of individuals promoting sustainability initiatives, and instructional materials enabling customers to make more sustainable decisions are a few examples of what this might contain. Furthermore, companies can design interactive experiences that let consumers participate with the brand's mission and meet other like-minded people, such workshops, virtual tours, or community gatherings. By giving customers a sense of being heard and appreciated, promoting user-generated content and comments can help to deepen the relationship. Through genuine brand alignment with environmental principles and the provision of opportunities for meaningful consumer interaction, environmentally conscious businesses may foster a devoted clientele that is committed to their sustainable goals.
3. *Inspiring exploration.* Brands that use eco-friendly products can use innovative marketing techniques to immerse consumers in the brand's sustainable goal, thereby creating an inspiring exploration for customers. One strategy is to create interactive experiences that inform and involve consumers on the environmental impact of their decisions, including pop-up exhibits, virtual reality tours, or hands-on workshops. These encounters can demonstrate the brand's environmentally friendly products in use, draw attention to their sustainable sourcing

and production methods, and offer helpful advice for leading a more environmentally conscious lifestyle.

4. *Evoking pride.* In order to instil a feeling of satisfaction in consumers for making environmentally conscious purchases, firms should utilize innovative tactics that highlight and validate the beneficial effects of their decisions. One strategy is to offer concrete proof of the product's sustainable credentials, including measurements on resource savings, reduced carbon emissions, or certifications. Customers are able to feel proud of themselves for choosing a morally sound choice because of this transparency. Additionally, through rewards programs, user-generated content platforms, or special events, corporations may foster a sense of community around eco-consciousness by praising and acknowledging consumers for their dedication to sustainable living.
5. *Impacting society.* Through the use of eco-friendly products, brands can have a positive social influence and promote shared responsibility by implementing innovative methods that encourage a sense of common purpose. Creating partnerships with non-profits, local communities, or environmental initiatives that enable consumers to make direct contributions to these causes through their purchases is one successful strategy. For instance, a percentage of profits could be given to programs that promote clean water, wildlife protection, or reforestation. Furthermore, companies have the power to inspire and enable consumers to take part in neighborhood clean-ups, recycling campaigns, and urban gardening initiatives, among other community-driven sustainability

What's Next?

When green marketing activities have successfully acquired the customer, the next step is to keep them and increase their frequency of buying the product and basket size. These three strategies will ensure that the customer will become a loyal one.

First, the consistency of conducting innovative green marketing practices. Numerous companies that produce environmentally friendly products use a variety of green marketing techniques to highlight their sustainable products and draw in eco-aware customers. Green product positioning is a useful tactic whereby companies emphasize the advantages and eco-friendliness of their goods in marketing materials and packaging. Businesses like Method encourage the use of natural, plant-based products and use eco-friendly packaging manufactured from recycled plastics. Certain firms employ green pricing methods, charging more for their environmentally friendly products, and make an effort to communicate their higher quality and sustainability. As part of their integrated green marketing strategies, many businesses also collaborate with environmental organizations, donate a portion of their profits to assist environmental issues, and use eco-friendly materials in product samples and promotional products.

Second, consistency in providing clear and helpful information to the customer in diverse channels to help them gain clarity and increase their

confidence in buying the product. Using digital channels like email marketing, social media, and websites to disseminate comprehensive information about the company's sustainability policies, eco-friendly materials, and product advantages is one successful strategy. Producing instructional materials like blog entries, infographics, and films can assist clients in comprehending the brand's dedication to sustainability and the advantages of their purchases. Furthermore, companies can use product packaging and in-store displays to promote sustainability claims, eco-certifications, and clearly labeled products that emphasize their green features. Collaborating with influencers, environmental advocacy groups, or outside promoters can further enhance the legitimacy and bolster the brand's environmentally conscious messaging. Eco-product businesses may address customer concerns, foster trust, and enable consumers to make confident, informed decisions about their purchases that are consistent with their environmental values by regularly providing clear and verifiable information across a variety of touchpoints.

Third, engage the loyal customer after they purchase the product. Businesses that sell environmentally friendly goods have a number of options for controlling consumer behavior after the sale and cultivating brand loyalty. One strategy is to offer thorough instructions for product upkeep and care, as well as pointers on extending the product's life and sustainability advantages. Additionally, brands can encourage customers to provide feedback and experiences, which can be used to enhance goods and services and create a community of environmentally conscious buyers. To further encourage sustainable practices, a rewards program that offers incentives for referrals, repeat business, or involvement in recycling campaigns might be put in place. Businesses can also educate consumers by showcasing the environmental impact of their products and encouraging sustainable lifestyle choices through webinars, blogs, or social media campaigns. Eco-friendly firms may build enduring relationships with customers and promote sustainable growth by providing outstanding post-purchase assistance, promoting product stewardship, and creating a feeling of community around shared environmental values.

Conclusion

In today's environmentally sensitive market, developing a successful green marketing plan is essential for companies looking to attract and keep eco-conscious consumers. Consumers actively seek out brands that share their values of sustainability and responsibility as they become more aware of their own environmental impact. A well-implemented green marketing plan fosters client trust, trustworthiness, and emotional ties in addition to showcasing a business's dedication to environmentally responsible activities. Businesses may stand out from rivals and strike a deep chord with customers who value environmental stewardship by openly disclosing sustainability programs, promoting cutting-edge green products, and encouraging a feeling of community purpose. Furthermore, long-term advocacy and loyalty can be fostered by green marketing tactics that inform and enable customers to make sustainable decisions. This will eventually enhance customer acquisition and retention. Creating a strong green marketing strategy is now essential for businesses to succeed and help ensure a more sustainable future in a time when environmental issues are gaining prominence.

References

Al Zubaidi, N. (2020). The relationship between collectivism and green product purchase intention: The role of attitude, subjective norms, and willingness to pay a premium. *Journal of Sustainable Marketing, 1*(1), 34–46.

Andrady, A. L. (2015). Persistence of plastic litter in the oceans. *Marine anthropogenic litter* (pp. 57–72).

Apte, S., & Sheth, J. N. (2016). *The sustainability edge: How to drive top-line growth with triple-bottom-line thinking*. University of Toronto Press.

Banyte, J., Brazioniene, L., & Gadeikiene, A. (2015). Communication of green marketing strategies for creating consumer awareness of green products. *Environmental Engineering and Management Journal, 14*(3), 671–682. Retrieved from https://www.diva-portal.org/smash/get/diva2:1119044/FULLTEXT01.pdf

Barbarossa, C., De Pelsmacker, P., & Moons, I. (2017). Personal values, green self-identity and electric car adoption. *Ecological Economics, 140*, 190–200. https://doi.org/10.1016/j.ecolecon.2017.05.015

Bhattacharya, C. B., & Polman, P. (2016). Sustainability lessons from the front lines. *MIT Sloan Management Review*.

Borchardt, M., Wendt, M. H., Pereira, G. M., & Sellitto, M. A. (2011). Redesign of a component based on ecodesign practices: Environmental impact and cost reduction achievements. *Journal of Cleaner Production, 19*, 49–57.

Chen, Y. S., & Chang, C. H. (2012). Enhance green purchase intentions: The roles of green perceived value, green perceived risk, and green trust. *Management Decision, 50*(3), 502–520.

Curkovic, S., Vickery, S., & Dröge, C. (2000). Quality-related action programs: Their impact on quality performance and firm performance. *Decision Sciences, 31*(4), 885–902.

D'Souza, C., Taghian, M., & Lamb, P. (2006). An empirical study on the influence of environmental labels on consumers. *Corporate Communications: An International Journal, 11*(2), 162–173. https://doi.org/10.1108/13563280610661697

Dangelico, R. M., & Vocalelli, D. (2017). "Green Marketing": An analysis of definitions, strategy steps, and tools through a systematic review of the literature. *Journal of Cleaner Production, 165*, 1263–1279.

de Jong, M. D., Harkink, K. M., & Barth, S. (2018). Making green stuff? Effects of corporate greenwashing on consumers. *Journal of Business and Technical Communication, 32*(1), 77–112.

Dehghanan, H., & Bakhshandeh, G. (2014). The impact of green perceived value and green perceived risk on green purchase behavior of Iranian consumers. *International Journal of Management and Humanity Sciences, 3*(2), 1349–1357.

Ganapathi, R., & Mahesh, N. (2013). Consumer's perceived value, attitude, and purchase intention of green products. *Management Insight, 9*(1), 36–43.

Geyer, R., Jambeck, J. R., & Law, K. L. (2017). Production, use, and fate of all plastics ever made. *Science Advances, 3*(7), e1700782.

Ginsberg-P, J. M., & Bloom, N. (2004). Choosing the right green marketing strategy. *MIT Sloan Management Review, 46*, 79.

Grankvist, G., Lekedal, H., & Marmendal, M. (2007). Values and eco-and fair-trade labelled products. *British Food Journal, 109*(2), 169–181.

Guo, R., Tao, L., Li, C. B., & Wang, T. (2017). A path analysis of greenwashing in a trust crisis among Chinese energy companies: The role of brand legitimacy and brand loyalty. *Journal of Business Ethics, 140*(3), 523–536.

Heikkurinen, P., Young, C. W., & Morgan, E. (2019). Business for sustainable change: Extending eco-efficiency and eco-sufficiency strategies to consumers. *Journal of Cleaner Production, 218*, 656–664.

Heimlich, J. E., & Ardoin, N. M. (2008). Understanding behavior to understand behavior change: A literature review. *Environmental Education Research, 14*(3), 215–237.

Herman, L., & Herman, E. (2014). Green marketing: Fostering the customers' willingness to buy green product in Indonesia. *International Journal of Data and Network Science, 6*(4), 1065–1076.

Hossain, M. I., & Rahman, M. S. (2018). Measuring influence of green promotion on green purchase behavior of consumers. *A Study on Bangladesh, 5*, 191–222.

International Energy Agency. (2020). Google Sustainability, n.d.; Microsoft Sustainability, n.d.; Apple Environmental Progress Report, n.d.

Jones, P., Comfort, D., & Hillier, D. (2011). Sustainability in the global shop window. *International Journal of Retail and Distribution Management, 39*(4), 256–271.

Joshi, Y., & Rahman, Z. (2015). Factors affecting green purchase behaviour and future research directions. *International Strategic Management Review, 3*(1-2), 128–143.

Jung, Y. J., & Kim, Y. (2023). Research trends of sustainability and marketing research, 2010–2020: Topic modeling analysis. *Heliyon, 9*(3).

Korindo Foundation. (2021). *Green sponsorship 2021*. Retrieved from https://korindofoundation.com/en/greensponsorship2021/

Krämer, L. (2020). Planning for climate and the environment: The EU green deal. *Journal for European Environmental and Planning Law, 17*(3), 267–306.

Kumar, P. (2016). State of green marketing research over 25 years (1990-2014): Literature survey and classification. *Marketing Intelligence and Planning, 34*(1), 137–158.

Kumar, V., & Christodoulopoulou, A. (2014). Sustainability and branding: An integrated perspective. *Industrial Marketing Management, 43*(1), 6–15.

Kumar, P., & Ghodeswar, B. M. (2015). Factors affecting consumers' green product purchase decisions. *Marketing Intelligence and Planning, 33*(3), 330–347. https://doi.org/10.1108/MIP-03-2014-0068

Lacy, P., & Rutqvist, J. (2015). *Waste to wealth: The circular economy advantage* (Vol. 91). Palgrave Macmillan.

Laroche, M., Bergeron, J., & Barbaro-Forleo, G. (2001). Targeting consumers who are willing to pay more for environmentally friendly products. *Journal of Consumer Marketing, 18*(6), 503–520.

Leonidou, L. C., Leonidou, C. N., Palihawadana, D., & Hultman, M. (2011). Evaluating the green advertising practices of international firms: A trend analysis. *International Marketing Review, 28*(1), 6–33.

Liao, Y.-K., Wu, W.-Y., & Pham, T.-T. (2020). Examining the moderating effects of green marketing and green psychological benefits on customers' green attitude, value and purchase intention. *Sustainability, 12*(18), 7461. https://doi.org/10.3390/su12187461

Lin, B., & Huang, H. (2012). The influence of green product competitiveness on the success of green product innovation: Empirical evidence from the Chinese electrical and electronics industry. *European Journal of Innovation Management, 15*(4), 468–490. https://doi.org/10.1108/14601061211127238

Little, V. J., Lee, C. K. C., & Nair, S. (2019). Macro-demarketing: The key to unlocking unsustainable production and consumption systems? *Journal of Macromarketing, 39*(2), 166–187.

Lofthouse, V. A., Bhamra, T. A., & Trimingham, R. L. (2009). Investigating customer perceptions of refillable packaging and assessing business drivers and barriers to their use. *Packaging Technology and Science, 22*(6), 335–348.

Magnier, L., & Schoormans, J. (2015). Consumer reactions to sustainable packaging: The interplay of visual appearance, verbal claim and environmental concern. *Journal of Environmental Psychology, 44*, 53–62.
Manongko, A. A. C., & Kambey, J. (2018). *The influence of green marketing on decision purchasing organic products with interests of buying as an intervening variable at Manado City, Indonesia*.
Martinho, G., Pires, A., Portela, G., & Fonseca, M. (2015). Factors affecting consumers' choices concerning sustainable packaging during product purchase and recycling. *Resources, Conservation and Recycling, 103*, 58–68.
Mehraj, D., & Qureshi, I. H. (2020). Determinants of green marketing mix in developing economies: Conceptualisation and scale validation approach. *Business Strategy and Development, 3*(4), 522–530.
Ottman, J. A. (2011). *The new rules of green marketing: Strategies, tools, and inspiration for sustainable branding*. Berrett-Koehler Publishers.
Parguel, B., Benoît-Moreau, F., & Larceneux, F. (2011). How sustainability ratings might deter 'greenwashing': A closer look at ethical corporate communication. *Journal of Business Ethics, 102*, 15–28.
Paul, J., Modi, A., & Patel, J. (2016). Predicting green product consumption using theory of planned behavior and reasoned action. *Journal of Retailing and Consumer Services, 29*, 123–134.
Peattie, K., & Peattie, S. (2009). Social marketing: A pathway to consumption reduction? *Journal of Business Research, 62*(2), 260–268.
Polonsky, M. J., & Rosenberger, P. J., III. (2001). Reevaluating green marketing: A strategic approach. *Business Horizons, 44*(5), 21–30.
Pretty, J. N., Ball, A. S., Lang, T., & Morison, J. I. L. (2005). Farm costs and food miles: An assessment of the full cost of the UK weekly food basket. *Food Policy, 30*(1), 1–19.
Prothero, A., Peattie, K., & McDonagh, P. (1997). Communicating greener strategies: A study of on-pack communication. *Business Strategy and the Environment, 6*(2), 74–82.
Pulido-Fernández, J. I., & López-Sánchez, Y. (2016). Are tourists really willing to pay more for sustainable destinations? *Sustainability, 8*(12), 1240.
Rahahleh, A. H., Moflih, M. A., Alabaddi, Z. A., Farajat, J., & Nawaf, S. (2020). The impact of green marketing on green consumer behavior in Jordan. *International Journal of Business and Management, 15*(1), 36–48.
Rahman, M. M., Siburan, R., & Noorlitaria, A. (2017). Green promotion indicators. In *Green promotion memediasi green packaging terhadap repurchase intention* (pp. 7500–7527). E-Jurnal Manajemen Unud.
Rahmat, T., Hurriyati, R., & Dirgantari, P. D. (2023). SDG's and zero emission vision in Indonesia: implementation of green marketing and 'green' direct marketing campaign opportunities based on population database. *International Journal of New Innovations, 1*(1), 1–15. Retrieved from https://journal.uinsgd.ac.id/index.php/ijni/article/view/22263
Rawlins, B. L. (2006). *Prioritizing stakeholders for public relations*. Institute for Public Relations. Retrieved from http://www.instituteforpr.org/research_single/prioritizing_stakeholders
Rejeki, S., Fauzi, A., & Yulianto, A. (2015). Green promotion or environmentally friendly promotion. *Dinasti International Journal of Digital Business Management, 3*(6), 1027.
Rokka, J., & Uusitalo, L. (2008). Preference for green packaging in consumer product choices – Do consumers care? *International Journal of Consumer Studies, 32*(5), 516–525.
Saravanan, A. P., Mathimani, T., Deviram, G., Rajendran, K., & Pugazhendhi, A. (2018). Biofuel policy in India: A review of policy barriers in sustainable marketing of biofuel. *Journal of Cleaner Production, 193*, 734–747.
Schmidheiny, S. (1992). *Changing course: A global business perspective on development and the environment* (Vol. 1). MIT Press.
Scott, K., Martin, D. M., & Schouten, J. W. (2014). Marketing and the new materialism. *Journal of Macromarketing, 34*(3), 282–290.
Seeger, M. W. (1997). *Ethics and organizational communication*. Hampton Press.

Sharma, A., & Iyer, G. R. (2012). Resource-constrained product development: Implications for green marketing and green supply chains. *Industrial Marketing Management, 41*(4), 599–608.

Sheth, J. N., & Parvatiyar, A. (2021). Sustainable marketing: Market-driving, not market-driven. *Journal of Macromarketing, 41*(1), 150–165.

Srisathan, W. A., Wongsaichia, S., Gebsombut, N., Naruetharadhol, P., & Ketkaew, C. (2023). The green-awakening customer attitudes towards buying green products on an online platform in Thailand: The multigroup moderation effects of age, gender, and income. *Sustainability, 15*(3), 2497.

Stengel, J. (2011). *Grow: How ideals power growth and profit at the world's greatest companies*. Crown Business.

Taufique, K. M. R., Vocino, A., & Polonsky, M. J. (2019). The influence of eco-label knowledge and trust on pro-environmental consumer behaviour in an emerging market. *Journal of Strategic Marketing, 27*(7), 604–620.

Testa, F., Iraldo, F., Vaccari, A., & Ferrari, E. (2015). Why eco-labels can be effective marketing tools: Evidence from a study on Italian consumers. *Business Strategy and the Environment, 24*(4), 252–265.

Thøgersen, J., Haugaard, P., & Olesen, A. (2010). Consumer responses to ecolabels. *European Journal of Marketing, 44*(11/12), 1787–1810.

Tiwari, R., Tripathi, R., & Srivastava, M. (2011). Green promotion strategies. *Journal of Business and Technical Communication, 32*(1), 77–112.

Tunley Environmental. (2023). *Social media and green marketing*. Retrieved from https://www.tunley-environmental.com/en/insights/the-power-of-social-media-in-green-marketing

Ummar, R., Shaheen, K., Bashir, I., Ul Haq, J., & Bonn, M. A. (2023). Green social media campaigns: Influencing consumers' attitudes and behaviors. *Sustainability, 15*(17), 12932. https://doi.org/10.3390/su151712932

Vatan, A., & Yilmaz, Z. (2020). New ceramic solutions in sustainable hotels within the scope of sustainable innovation. In *Entrepreneurial opportunities* (pp. 147–163). Emerald Publishing Limited.

Verma, V. K., Chandra, B., & Kumar, S. (2019). Values and ascribed responsibility to predict consumers' attitude and concern towards green hotel visit intention. *Journal of Business Research, 96*, 206–216.

Villena, V. H., & Gioia, D. A. (2020). A more sustainable supply chain. *Harvard Business Review, 98*(2), 84–93.

Wang, L., & Wong, P. P. W. (2019). The demographic impact of consumer green purchase intention: A study of environmental conscious consumers. *International Journal of Tourism Sciences, 19*(4), 269–281. https://doi.org/10.1080/15980634.2019.1686594

Widlitz, S. (2020). Retailers get serious about sustainability into 2020. *Forbes*.

Wijen, F., Zoeteman, K., Pieters, J., & van Seters, P. (2012). *A handbook of globalisation and environmental policy: National government interventions in a global arena*. Edward Elgar Publishing.

Winston, N., Kennedy, P., & Carlow, I. (2019). *Severe housing deprivation: Addressing the social sustainability challenge in the EU* (Vol. No. 201903). Geary Institute, University College Dublin.

Yunus, M., & Desai, C. P. (2017). *Perception of scientists of Anand Agricultural University towards organizational climate*.

Zhang, L., Xu, Y., Oosterveer, P., & Mol, A. P. J. (2023). Enhancing green purchase intentions: The effects of product transformation salience and consumer traceability knowledge. *Journal of Business Ethics*. https://doi.org/10.1007/s10551-023-05555-1

Dina Dellyana is currently an assistant professor in innovation management and business incubator coordinator at the School of Business and Management, Institut Teknologi Bandung (SBM ITB). She's also a managing partner and founder of Innovative DNA Solution, a research and consulting company focus in innovation and business strategy. She's actively conducting research with topics on Creative Industry, Digitalization, Innovation, SME Development, and Business Incubator. She writes several book and book chapters such as Blueprint Book of Indonesia's Creative Industry: Music Subsector for Ministry of Tourism and Creative Economy (2015), Routledge Handbook of Cultural and Creative Industry in Asia" (2018), Collage of National Creative Economy Thought (2018), Digital Incubator Playbook (2020), Modelling a Healthy Creative Scene (2021), Startup Tools Book (2021), and Childpreneurship (2022). Since 2014, she has been conducting research and consulting on creative economy.

Leo Aldianto graduated from Doctor of Science Management (DSM) from Institut Teknologi Bandung, Master's degree in Aerospace Engineering from Delft University of Technology in the Netherlands in 1993 and a Master of Business Administration (MBA) from the TSM Business School the Netherlands in 2000. After working in the Aerospace and IT industry for 14 years, Leo joined the Entrepreneurship and Technology Management interest group in SBM, with expertise/specialization in Management of Innovation, Entrepreneurship, New venture creation, and Product & Business Development.

Part II
An International View on Sustainable Business Development: Country Studies

Chapter 5
The Influence of Incentive Alignment, Cost Sharing, and Responsible Action Scenario to Agriculture Supply Chain Performance in an Indonesian Study (Dynamic Simulation Approach)

Irayanti Adriant, Togar Mangihut Simatupang, and Yuanita Handayati

Abstract This research investigates the dynamics of agricultural supply chains, specifically focusing on the broccoli supply chain, through a dynamic simulation approach. The study emphasizes the importance of incentive alignment, cost-sharing mechanisms and responsible actions in enhancing the performance and sustainability of agricultural practices. By examining various scenarios, the research highlights how cost-sharing can alleviate financial burdens on farmers, enabling them to invest in sustainable practices and improve their profitability. Additionally, the study explores the role of retailer incentives in motivating farmers to adopt better agricultural techniques, ultimately leading to higher quality produce and increased profits for all stakeholders involved. The findings underscore the necessity of collaboration and shared goals among supply chain actors to foster a more resilient and sustainable agricultural system, contributing to food security and environmental preservation.

Keywords Sustainable agriculture supply chain · Dynamic simulation · responsible action

---

I. Adriant (✉)
Logistics Management Department, Universitas Logistik dan Bisnis Internasional, Bandung, Indonesia
e-mail: Irayanti@ulbi.ac.id

T. M. Simatupang · Y. Handayati
School of Business and Management, Institut Technology Bandung, Bandung, Indonesia
e-mail: togar@sbm-itb.ac.id; yuanita@sbm-itb.ac.id

A. H. Verkuil et al. (eds.), *Core Values and Decision-Making for Sustainable Business*, Sustainable Business Development,
https://doi.org/10.1007/978-3-031-78361-6_5

Introduction

Sustainability in business is increasingly gaining attention as a critical approach that aims to achieve long-term profits while considering social, environmental, and economic impacts (Olwig, 2021). Companies focusing on sustainability are more likely to thrive, as they prioritize resource management and seek environmentally friendly solutions, reducing risks associated with fluctuating raw material prices and regulatory changes. Furthermore, businesses committed to sustainability attract customers and investors who prefer companies with strong social and environmental values, thus fostering growth and building long-term relationships. This commitment not only enhances a company's image but also fulfills corporate social responsibility (CSR) obligations, leading to greater support from society and government, and ultimately contributing to sustainable development (Coffay & Bocken, 2023; Maas & Reniers, 2014; Zhong et al., 2023).

The supply chain plays a vital role in modern business by managing and optimizing the flow of goods and services from producers to consumers (Burgess et al., 2023). An efficient supply chain reduces costs, increases efficiency, and maintains product quality, while also responding to customer demands in a fast-paced market. By effectively managing their supply chains, companies can avoid stock shortages, enhance customer satisfaction, and mitigate risks associated with supply disruptions (Chandiran et al., 2023). Thus, the supply chain is the backbone of modern business, essential not only for cost management and efficiency but also for maintaining competitiveness in an evolving marketplace.

A sustainable supply chain is an approach that emphasizes the importance of minimizing environmental and social impacts throughout the entire life cycle of a product or service. In an environmental context, sustainable supply chains seek to reduce carbon footprints, optimize the use of natural resources, and reduce waste. This can be achieved through the use of green technology, the selection of environmentally friendly raw materials and more efficient packaging strategies (Chaudhuri et al., 2023; Karmaker et al., 2023; Zheng & Zhang, 2023). In this way, companies that adopt sustainable supply chains can reduce negative impacts on the environment, while also responding to the demands of consumers who are increasingly concerned about environmental issues. In the social aspect, a sustainable supply chain takes into account the welfare of workers and local communities. This includes fair wages, safe working conditions, as well as corporate social responsibility in the surrounding community. Companies that care about social aspects in their supply chains are able to create a better work environment, build positive relationships with workers, and provide benefits to the surrounding community (Bubicz et al., 2019; Gupta et al., 2022). Sustainable supply chain also takes economic aspects into consideration. Although the main focus is on environmental and social aspects, business sustainability is also a key element. By minimizing waste, companies can reduce production costs, increase efficiency, and increase profitability. In the context of agriculture supply chain, sustainability are key to ensuring food security and a healthy environment (Jiang et al., 2022). Sustainability in agricultural supply chains includes wise management of natural resources. This includes practices such as the use of organic fertilizers, efficient irrigation, and sustainable land management to minimize soil erosion. Thus, sustainable agriculture can maintain long-term productivity while reducing negative impacts on the environment (Jiang et al., 2022).

A sustainable agricultural supply chain also takes into account the welfare of farmers and agricultural workers. This includes fair wages, safe working conditions, and training and education for farmers to improve their capabilities in sustainable agricultural practices (Handayati et al., 2015). By ensuring the welfare of farmers, agricultural supply chains can become more stable and sustainable. Sustainable agricultural supply chains also promote fair trade practices (Adriant et al., 2023). This includes strong partnerships between manufacturers, distributors and consumers that enable a more equitable sharing of benefits in the supply chain. Through fair trade, agricultural producers have greater opportunities to increase their income, while consumers gain access to quality products. By running sustainable supply chains in agriculture, we can maintain enough food for the global population while protecting the environment and ensuring the welfare of farmers. This is critical to meeting the demands of a growing world population and addressing climate change that is increasingly affecting agriculture (De Fazio, 2016).

Sustainable supply chains in agriculture are still difficult to implement, especially in developing countries, due to a number of complex challenges faced in this industry (Adriant et al., 2021). Some of the main reasons include: limited resources and technology (da Silveira et al., 2023), high food demands (Grabowski & Self, 2023; Raihan et al., 2023), dependence on Global Market (Malila et al., 2023; Martin et al., 2015), required investments (Assa et al., 2021; Jauhari et al., 2023), and awareness and education (Adriant et al., 2021). The implementation of sustainable supply chain were hardly implemented because of the absence of incentive alignment scheme. Incentive alignment, within the context of implementing sustainable agriculture supply chains, is a crucial concept (Simatupang et al., 2000). It refers to the need to ensure that all parties involved, from farmers and producers to distributors and retailers, share common goals and incentives that prioritize sustainability. Another factor that influences the implementation of sustainable supply chains in the agricultural sector is cost sharing activities. First, in an agricultural context, investing in sustainable practices such as water-saving irrigation, organic fertilization, or environmentally friendly pest control can require high initial costs. Cost sharing allows farmers, producers, and other parties in the supply chain to share this financial burden. Cost sharing also promotes collaboration and interdependence in agricultural supply chains. By sharing costs and benefits, all parties in the supply chain have greater motivation to maintain sustainability (Herbon & David, 2023; Zheng et al., 2022). This creates stronger relationships between farmers, producers, distributors and consumers, which in turn can increase the availability of sustainable agricultural products on the market and meet the demands of consumers who are increasingly concerned about sustainability.

This research will discuss the relationship between incentive alignment, cost sharing, and other responsible actions on the performance of sustainable agriculture supply chains, which in this case is measured by equal distribution of profits between supply chain actors. So far, farmers, as the most important part of the agricultural supply chain, have incomes that are not the same as other supply chain actors, such as middlemen or retailers. This research will discuss if the incentive alignment scenario and cost sharing scenario are implemented, how the behavior of the regional agri-supply chain system will change.

Methodology

The methodology section describes the approach taken in the research, focusing on dynamic systems simulation to analyze the agricultural supply chain. This methodology involves several key steps to ensure a comprehensive understanding of the system's behavior over time. The first step is identifying the system to be simulated, which in this case is the agricultural supply chain involving various actors such as farmers, middlemen, and retailers. Once the system is defined, the next phase involves building a mathematical model that represents the relationships and interactions among these actors. This model serves as the foundation for simulating the dynamics of the supply chain.

Determining initial conditions is crucial as it sets the starting parameters for the simulation. This includes factors like initial stock levels, production costs, and performance metrics for each actor in the supply chain. After establishing these conditions, the simulation is run to observe how the system evolves over time under different scenarios. Validation and verification are essential to ensure the accuracy and reliability of the model. This involves comparing the simulation output with actual data to confirm that the model accurately reflects real-world behavior. Statistical methods are employed to assess this alignment, ensuring that any discrepancies are addressed before proceeding with scenario analysis. The final step involves running various scenarios to explore how changes in variables, such as cost-sharing and incentive alignment, impact the performance of the agricultural supply chain. Each scenario is analyzed to understand its effects on key outcomes, such as farmer profits and overall system efficiency. This iterative process allows researchers to refine their understanding of the dynamics at play and develop actionable insights for improving sustainability and profitability in agricultural supply chains.

Result and Discussion

This research uses a broccoli supply chain as study case. The supply chain consists of several actors such as farmers, middlemen, and retailers. Each entity in the supply chain has a different function and role. The farmer as a producer will buy broccoli seeds from his supplier and will carry out the planting process. After harvesting, the farmer will sell the crop to the farmer community. The farmer community will collect broccoli from several farmers and carry out the process of sorting and checking the quality of broccoli. The broccoli will then be packed and sent to retailers. Retailers will sell the broccoli to end consumers.

Business as Usual

Currently, farmers need to pay high costs for broccoli production materials such as seeds, pesticides, fertilizers, mulch, lime, land, water, and the farmer's own labor. In addition, the profit obtained by farmers is the lowest compared to Gapoktan and retailers. In addition, there is no cooperation between each actor in supporting the welfare of farmers to maintain national food security, especially for broccoli. The application of technology both for production such as plowing machines or harvesting machines, and as an information system is expected to increase the productivity and profits of farmers. Cooperation between each actor is needed to invest in agricultural technology by taking some percentage of the profits of Gapoktan and retailers. In addition to the application of agricultural technology, the application of the Responsible Supply Chain (RSC) action is also expected to increase farmers' profits, and even increase profits from Gapoktan and retailers. The implementation of the RSC action can also improve the image of each actor of the broccoli supply chain in Cikidang Village, Lembang, because each actor cooperates in technology investment for equal benefits obtained from each actor of the broccoli supply chain.

The broccoli supply chain system has 10 loops (reinforcement and balancing loops). From a causal loop diagram, we analysed stock variables and flows that occur in the system. This system consist of several sub models such as a sub model of farmer's profit, of output, of production costs, of middlemen's (Gapoktan's) profit and retailer's profit.

Model Validation and Verification

The model was verified to ensure that it contains no errors. The authors built a model in the Anylogic 7.0.2 application. Model validation was conducted using statistical validation which compares the output from a simulation and the actual data and calculates the Mean Absolute Percentage Error (MAPE).

The MAPE analysis gives the following data:

Day	At Historical data	Ft Output model	At - Ft	|At - Ft|	|(At - Ft)/At|
1	−71.110,000	−71.110,000	0	0	0
30	−71.110,000	−71.110,000	0	0	0
56	−66.500.000	−66,502,000	2000	2000	0.00000301

Day	At Historical data	Ft Output model	At - Ft	|At - Ft|	|(At - Ft)/At|
71	−41,930,000	−41,925,616	−4.384	4.384	0.0001046
91	−14,280,000	−14,277,616	−2384	2384	0.0001670
120	33,340,000	33,338,768	1.232	1.232	0.00000369
Total					0.0057280
Number of data					120
Mean absolute percentage error					0.0047734

Based on these results, the value of the Mean Absolute Percentage Error on the Profit obtained by farmers is 0.00477%. From this value, it can be interpreted that the ability of the model is valid.

Model's Scenarios

There are five scenarios that will be simulated.

1. Scenario 1: Implementation of cost sharing.

   In the first scenario, the implementation of cost sharing is examined. This approach involves a Gapoktan (a cooperative of a group of farmers) assisting farmers by sharing the costs associated with broccoli production. By reducing the financial burden on farmers, who typically incur high costs for seeds, fertilizers, and labor, this scenario aims to increase their income. The rationale behind this scenario is that when costs are shared, farmers can invest more in their operations, leading to improved productivity and profitability. The expected impact is a significant increase in farmers' profits, as they would only need to bear half of the production costs, thus allowing for greater financial stability and encouraging sustainable practices.

   Gapoktan assistance to farmers is a way to help farmers by sharing the incurred costs. The implementation of this assistance is expected to increase farmers' income, because the incurred costs are not only borne by the farmers but also by the Gapoktan. The cost of farmer is Rp 8000/Kg by using this scenario, farmers will only bear Rp 4000/Kg as the cost of broccoli. The system performance is the farmer profit.
2. Scenario 2: Implementation of incentive alignment by retailer.

   The second scenario focuses on incentive alignment through retailer involvement. In this scenario, retailers provide incentives to farmers based on their performance. Specifically, retailers offer rewards to farmers who meet or exceed a minimum performance threshold. This incentive structure is intended to motivate farmers to adopt better agricultural practices, resulting in higher quality produce. The rationale here is that when farmers are rewarded for their efforts, they are more likely to invest in improving their farming techniques, which ultimately benefits the entire supply chain. The potential impact includes enhanced product quality and increased profits for both farmers and retailers, fostering a more collaborative and productive relationship within the supply chain.

Retail incentives help retailers to benefit and to increase their profits. Retail incentives can have the form of rewards for the farmers or be related to their CSR management. Farmers will receive a reward upon demonstrating performance that meets or exceeds the established minimum standards. This reward is expected to motivate farmers to plant well so the retail will get good results. The second incentive are CSR-related funds. The assumption is that fulfilling sustainability requirements expected by retailers can increase the sale. So both sides will benefit.

3. scenario 3: Implementation of incentive alignment by government.

   Third scenario focuses on the incentive alignment given by the government. The government provides incentives and facilities for horticultural businesses. In the government regulation there are financial assistance, tax cuts, promotions to improve the image of the business and etc. This research assumes three incentives. The first incentive is financial assistance to a Gapoktan of Rp 840,000/day or the total financial assistance for 4 months is Rp 100,800,000. The second incentive is to provide a tax discount of 10% to retail. The third incentive is to do promotions and it is assumed to increase sales to end consumers by 40 kg/3 days.

4. scenario 4: Implementation of other responsible action

   The fourth scenario focuses on responsible action that consists of three indexes: the first is the health index, which is represented by how many farmers have a health insurance. The second is the welfare index, which is represented by working hours and training hours of supply chain actors. The third describes the use of pesticides and the circular economy. The assumption is that if the RSC is carried out, it will have an impact on two outputs, those are an increase in the selling price of broccoli from each actor and an increase in sales to end consumers.

5. scenario 5: Combined scenario

   This scenario highlights the importance of collaboration and shared goals among supply chain actors. This scenario is the combination of scenario 1, 2,3 and 4. This scenario will show the system behavior if all the scenarios activated.

From the simulation result, it shows that the largest increase in farmer profits is in the scenario of cost sharing + government incentives + retail incentives, with an increase of 171% from the existing scenario. On the other hand, the biggest loss for farmers is in the crop failure scenario, which decreases by 91%. The biggest profit obtained by middlemen (Gapoktan) is in the scenario of government incentives + retail incentives + RSC, with an increase of 525%. Meanwhile, the biggest loss from middleman (Gapoktan) is in the cost sharing scenario which gets a decrease of 85%. Furthermore, the biggest profit increase of 48% is obtained by retail in the scenario of cost sharing + government incentives + RSC.

The results of this study state that collaboration between actors in the responsible agriculture supply chain must be carried out simultaneously. The role of all actors is very influential on the successful implementation of this responsible agriculture supply chain. On the retailer side, many retailers in Indonesia still act transactionally, where they will not carry out any activity if there is no benefit to be received.

To overcome this, the scenario of government involvement by providing incentives for retailers if they carry out responsible actions will make retailers feel they have more profit. The average supply chain profit and the average profit of each member in supply chain based on the result of simulation shown in Table 5.1.

The findings from the research on agricultural supply chains, particularly in the context of broccoli production, suggest that implementing cost-sharing and incentive alignment can significantly enhance the performance and sustainability of these systems. However, there are potential counterarguments and challenges to consider.

One counterargument to the effectiveness of cost-sharing is the financial strain it may place on organizations like Gapoktan. While sharing costs can reduce the immediate financial burden on farmers, it may also lead to increased operational costs for the supporting organizations. If these organizations are unable to sustain their financial health, the long-term viability of the cost-sharing model could be jeopardized. Additionally, there may be concerns regarding the equitable distribution of costs and benefits among all parties involved. If not managed properly, cost-sharing could result in conflicts or dissatisfaction among stakeholders, particularly if some farmers benefit more than others.

In terms of incentive alignment, while providing rewards to farmers based on performance can motivate better agricultural practices, it also raises questions about the feasibility of establishing fair and measurable performance metrics. If the criteria for performance are perceived as too stringent or unrealistic, farmers may feel discouraged rather than motivated. Additionally, the reliance on retailer incentives can create a dependency that may not be sustainable in the long run, especially if market conditions change or if retailers decide to withdraw their support. Moreover, both scenarios assume that all actors within the supply chain are willing and able to collaborate effectively. In reality, there may be significant barriers to cooperation, such as differing priorities, lack of trust, or inadequate communication among stakeholders. These barriers can hinder the successful implementation of cost-sharing and incentive alignment strategies, potentially leading to suboptimal outcomes.

Finally, while the research highlights the potential benefits of these strategies, it is essential to consider the broader economic and environmental context. For instance, external factors such as market fluctuations, climate change, and

Table 5.1 Simulation Result comparation

Scenario	Farmer profit	Middleman (Gapoktan) profit	Retailer profit	Average of Supply chain profit	BEP (day)
BAU/ Eksisting	Rp33.338.768	Rp40.526.578	Rp168.940.868	Rp242.806.214	98
Scenario 1	Rp62.270.000	Rp6.273.000	Rp168.900.000	Rp237.443.000	89
Scenario 2	Rp49.660.000	Rp83.960.000	Rp135.300.000	Rp268.920.000	95
Scenario 3	Rp31.020.000	Rp163.300.000	Rp186.200.000	Rp380.520.000	101
Scenario 4	Rp46.210.000	Rp97.960.000	Rp228.800.000	Rp372.970.000	95
Scenario 5	Rp83.920.000	Rp172.600.000	Rp147.900.000	Rp404.420.000	86

regulatory changes can impact the effectiveness of cost-sharing and incentive alignment. If these external factors are not adequately addressed, the intended benefits of improved supply chain performance may not be realized. In summary, while the findings suggest promising pathways for enhancing agricultural supply chains through cost-sharing and incentive alignment, it is crucial to remain aware of potential counterarguments and challenges. Addressing these concerns through careful planning, stakeholder engagement, and adaptive strategies will be essential for realizing the full potential of these approaches in building sustainable agricultural supply chains.

Conclusion

The research highlights the significant impact of incentive alignment, cost sharing, and responsible actions on the performance of agricultural supply chains, particularly in the context of broccoli production. Key findings indicate that implementing cost-sharing mechanisms can alleviate the financial burden on farmers, allowing them to invest more in sustainable practices and ultimately increasing their profitability. This approach not only supports the economic stability of farmers but also fosters collaboration among supply chain actors, enhancing overall system resilience. In addition, the study emphasizes the importance of incentive alignment, particularly through retailer involvement. By offering performance-based rewards, retailers can motivate farmers to adopt better agricultural practices, leading to improved product quality and higher profits for both farmers and retailers. This alignment of interests is crucial for creating a more cooperative and productive supply chain, which can adapt more effectively to market demands and environmental challenges. The implications of these findings are profound. They suggest that fostering collaboration and shared goals among supply chain participants can lead to more sustainable agricultural practices, ultimately contributing to food security and environmental preservation. Policymakers and industry stakeholders are encouraged to consider these strategies when designing interventions to enhance agricultural supply chains. By focusing on cost sharing and incentive alignment, the agricultural sector can move towards a more sustainable future, benefiting not only the farmers but also the entire supply chain and society at large.

References

Adriant, I., Simatupang, T. M., & Handayati, Y. (2021). The barriers of responsible agriculture supply chain: The relationship between organization capabilities, external actor involvement, and supply chain integration. *Uncertain Supply Chain Management, 9*(2), 403–412. https://doi.org/10.5267/j.uscm.2021.2.003

Adriant, I., Simatupang, T. M., & Handayati, Y. (2023). Collaboration in responsible agriculture supply chain management: A systematic literature review. *International Journal of Integrated Supply Management, 16*(2), 148–173. https://doi.org/10.1504/IJISM.2023.130327

Assa, H., Sharifi, H., & Lyons, A. (2021). An examination of the role of price insurance products in stimulating investment in agriculture supply chains for sustained productivity. *European Journal of Operational Research, 288*(3), 918–934. https://doi.org/10.1016/j.ejor.2020.06.030

Bubicz, M. E., Barbosa-Póvoa, A. P. F. D., & Carvalho, A. (2019). Incorporating social aspects in sustainable supply chains: Trends and future directions. *Journal of Cleaner Production, 237*. https://doi.org/10.1016/j.jclepro.2019.06.331

Burgess, P. R., Sunmola, F. T., & Wertheim-Heck, S. (2023). A review of supply chain quality management practices in sustainable food networks. *Heliyon, 9*(11), e21179. https://doi.org/10.1016/j.heliyon.2023.e21179

Chandiran, P., Ramasubramaniam, M., Venkatesh, V. G., Mani, V., & Shi, Y. (2023). Can driver supply disruption alleviate driver shortages? A systems approach. *Transport Policy, 130*(October 2022), 116–129. https://doi.org/10.1016/j.tranpol.2022.10.002

Chaudhuri, R., Chatterjee, S., Gupta, S., & Kamble, S. (2023). Green supply chain technology and organization performance: Moderating role of environmental dynamism and product-service innovation capability. *Technovation, 128*(August), 102857. https://doi.org/10.1016/j.technovation.2023.102857

Coffay, M., & Bocken, N. (2023). Sustainable by design: An organizational design tool for sustainable business model innovation. *Journal of Cleaner Production, 427*(September), 139294. https://doi.org/10.1016/j.jclepro.2023.139294

da Silveira, F., Barbedo, J. G. A., da Silva, S. L. C., & Amaral, F. G. (2023). Proposal for a framework to manage the barriers that hinder the development of agriculture 4.0 in the agricultural production chain. *Computers and Electronics in Agriculture, 214*(August), 108281. https://doi.org/10.1016/j.compag.2023.108281

De Fazio, M. (2016). Agriculture and sustainability of the welfare: The role of the short supply chain. *Agriculture and Agricultural Science Procedia*. https://doi.org/10.1016/j.aaspro.2016.02.044

Grabowski, R., & Self, S. (2023). Agricultural productivity growth and the development of manufacturing in developing Asia. *Economic Systems, 47*(2), 101075. https://doi.org/10.1016/j.ecosys.2023.101075

Gupta, P., Mehlawat, M. K., Aggarwal, U., & Khan, A. Z. (2022). An optimization model for a sustainable and socially beneficial four-stage supply chain. *Information Sciences, 594*, 371–399. https://doi.org/10.1016/j.ins.2022.02.032

Handayati, Y., Simatupang, T. M., & Perdana, T. (2015). Agri-food supply chain coordination: The state-of-the-art and recent developments. *Logistics Research, 8*(1), 1–15. https://doi.org/10.1007/s12159-015-0125-4

Herbon, A., & David, I. (2023). Optimal manufacturer's cost sharing ratio, shipping policy and production rate – A two-echelon supply chain. *Operations Research Perspectives, 10*(July 2022), 100264. https://doi.org/10.1016/j.orp.2022.100264

Jauhari, W. A., Novia Ramadhany, S. C., Nur Rosyidi, C., Mishra, U., & Hishamuddin, H. (2023). Pricing and green inventory decisions for a supply chain system with green investment and carbon tax regulation. *Journal of Cleaner Production, 425*(August), 138897. https://doi.org/10.1016/j.jclepro.2023.138897

Jiang, Y., Li, K., Chen, S., Fu, X., Feng, S., & Zhuang, Z. (2022). A sustainable agricultural supply chain considering substituting organic manure for chemical fertilizer. *Sustainable Production and Consumption, 29*, 432–446. https://doi.org/10.1016/j.spc.2021.10.025

Karmaker, C. L., Al Aziz, R., Ahmed, T., Misbauddin, S. M., & Moktadir, M. A. (2023). Impact of industry 4.0 technologies on sustainable supply chain performance: The mediating role of green supply chain management practices and circular economy. *Journal of Cleaner Production, 419*(June), 138249. https://doi.org/10.1016/j.jclepro.2023.138249

Maas, S., & Reniers, G. (2014). Development of a CSR model for practice: Connecting five inherent areas of sustainable business. *Journal of Cleaner Production, 64*, 104–114. https://doi.org/10.1016/j.jclepro.2013.07.039

Malila, B. P., Kaaya, O. E., Lusambo, L. P., Schaffner, U., & Kilawe, C. J. (2023). Factors influencing smallholder Farmer's willingness to adopt sustainable land management practices to control invasive plants in northern Tanzania. *Environmental and Sustainability Indicators, 19*(July 2022), 100284. https://doi.org/10.1016/j.indic.2023.100284

Martin, S., Rieple, A., Chang, J., Boniface, B., & Ahmed, A. (2015). Small farmers and sustainability: Institutional barriers to investment and innovation in the Malaysian palm oil industry in Sabah. *Journal of Rural Studies, 40*, 46–58. https://doi.org/10.1016/j.jrurstud.2015.06.002

Olwig, M. F. (2021). Sustainability superheroes? For-profit narratives of "doing good" in the era of the SDGs. *World Development, 142*. https://doi.org/10.1016/j.worlddev.2021.105427

Raihan, A., Ibrahim, S., & Muhtasim, D. A. (2023). Dynamic impacts of economic growth, energy use, tourism, and agricultural productivity on carbon dioxide emissions in Egypt. *World Development Sustainability, 2*(April), 100059. https://doi.org/10.1016/j.wds.2023.100059

Simatupang, T. M., Sridharan, R., Wright, A. C., & Zealand, N. (2000). Information and incentives in a supply chain. *The 7th annual New Zealand engineering and technology postgraduate conference*, *March*, 269–274.

Zheng, Y., & Zhang, Q. (2023). Digital transformation, corporate social responsibility and green technology innovation- based on empirical evidence of listed companies in China. *Journal of Cleaner Production, 424*(1550), 138805. https://doi.org/10.1016/j.jclepro.2023.138805

Zheng, B., Wen, K., Jin, L., & Hong, X. (2022). Alliance or cost-sharing? Recycling cooperation mode selection in a closed-loop supply chain. *Sustainable Production and Consumption, 32*, 942–955. https://doi.org/10.1016/j.spc.2022.06.001

Zhong, C., Guo, H., Swan, I., Gao, P., Yao, Q., & Li, H. (2023). Evaluating trends, profits, and risks of global cities in recent urban expansion for advancing sustainable development. *Habitat International, 138*(August 2022), 102869. https://doi.org/10.1016/j.habitatint.2023.102869

Irayanti Adriant, Dr., is a lecturer at the Faculty of Logistics, Technology, and Business, Universitas Logistik dan Bisnis Internasional, Bandung, Indonesia. She holds a doctoral degree in Operation Management from The School of Business and Management, Bandung Institute of Technology. Her research focuses on operation research, logistics and supply chain management, and sustainability. She teaches courses in supply chain modelling and simulation, quality management and operation management. Currently, she serves as the head of logistics Management Department at the university.

Togar Mangihut Simatupang, Prof., is a Professor of Operations and Supply Chain Management at the School of Business and Management in Bandung Institute of Technology, Indonesia. He has extensively published in logistics and supply chain management journals. He has been recipient of the Highly Commended Award by Emerald Literati Network for his research in supply chain management. His current research and teaching interests focus primarily on supply chain management, logistics systems, value chain management, creative economy, design thinking, innovation system, and entrepreneurship.

Yuanita Handayati, Dr., is a faculty member at the School of Business and Management, Institut Teknologi Bandung. Her research areas include agricultural supply chains, circular economy, and sustainability. She teaches courses in operations management, logistics systems, international transport, and supply chain management. Additionally, she has a strong interest in gamification and game-based learning. Currently, she serves as the Head of the Educational Games Laboratory and holds the position of Head of International Accreditation for the School.

Chapter 6
From Values to Practice: Evaluating the Intention of Indian MSMEs to Adopt Sustainable Business Practices

Minu Mehta and Boishampayan Chatterjee

Abstract There are around 634000 MSMEs in India in 2024, of which 90% come under the micro category. MSMEs account for 25% of the total energy consumed by industries in India and so are critical to India's attempts to adopt sustainable business practices. Of late, there has been a heightened interest in this sector towards the adoption of sustainable practices related to the environment, social and governance (ESG) parameters. There is a growing realization that of ESG alignment can be the next big factor offering positive differentiation to Indian MSMEs, in addition to low labour cost and demographic advantage. The authors found academic research centered around the obstacles or enablers that influence adoption of sustainable practices in MSME businesses and identified the need to study the factors that could influence the intention of MSMEs to adopt or reject pro-sustainability practices. A conceptual model was developed to understand the factors that influence the intention and through it, the decision-making behavior of MSME entrepreneurs and business persons. The model shows a two-tiered relationship, firstly, between the four core parameters, namely, Performance Benefits, Effort Assessment, Enabling Conditions and Social Pressure and intention, and secondly, between the influence of three moderating factors (Sector, Turnover and Location), on the four core parameters.

Keywords MSME · Sustainable practices · decision making behavior

M. Mehta (✉) · B. Chatterjee
Anil Surendra Modi School of Commerce, SVKM's Narsee Monjee Institute of Management Studies (NMIMS) Deemed-to-University, Mumbai, India
e-mail: minu.mehta@nmims.edu; boishampayan.chatterjee@nmims.edu

A. H. Verkuil et al. (eds.), *Core Values and Decision-Making for Sustainable Business*, Sustainable Business Development,
https://doi.org/10.1007/978-3-031-78361-6_6

Introduction

In 2024, India has approximately 634,000 MSMEs, with 90% classified as micro-enterprises.[1] These MSMEs consume about 25% of the energy used by industries in India,[2] making them crucial to the country's efforts to embrace sustainable business practices. Recently, there has been growing interest in this sector in adopting sustainable practices, particularly concerning environmental, social, and governance (ESG) parameters.[3] There is a growing realization that of ESG alignment can be the next big factor offering positive differentiation to Indian MSMEs, in addition to low labour cost and demographic advantage. Recognizing the growing market for ESG solutions, Bizongo, a supply chain digitizing platform, launched the ESG Scorecard,[4] in 2022, to help MSMEs to track their sustainability initiatives. Along similar lines, the Sustainability Perception Index[5] was launched by Small Industries Development Bank of India and Dun and Bradstreet to monitor the awareness, willingness and implementation of sustainability initiatives by MSMEs. This paper investigates the factors that influence the intention of MSMEs to adopt sustainable business practices. The authors use the Theory of Planned Behavior (Ajzen, 1991), as a template to develop a conceptual model that helps to highlight the parameters MSMEs use to evaluate the costs and benefits of integrating green practices and the specific functions or verticals picked up by them. Primary data collected from 41 MSME business owners with the help of in-depth interview and structured questionnaire offers insights for the relative importance of various factors, especially given the context of lack of clarity and a clear government mandate to go green. In fact, Abdullah et al. (2023), identified government and regulatory barriers as the second most significant barrier out of 21 that need to be overcome for implementing sustainable manufacturing practices. Basit et al. (2024) found that even though MSMEs struggle to integrate sustainable practices due to lack of financial and technological resources, they exhibit an agility towards adoption of change on account of simpler business processes. Nudurupati et al. (2022) report lack of cooperation from key stakeholders to be a major deterrent that hinders circular economy adoption by MSMEs and expectations of customers to be an important driver pushing for adoption. Kumar et al. (2022) opine that lack of resources and technical expertise are the two most critical barriers faced by MSMEs. According to Vásquez et al. (2024), the simpler the green technologies are, greater are the chances of MSMEs understanding, accepting and adopting them. The authors found academic research centred around the obstacles or enablers that influence adoption of sustainable practices in MSME businesses and identified the need to study the factors that could influence

---

[1] https://www.forbes.com/advisor/in/business/msme-statistics/

[2] https://indiacsr.in/esg-scorecard-to-help-msmes-to-achieve-compliance-and-sustainability/

[3] https://economictimes.indiatimes.com/small-biz/sme-sector/how-esg-can-become-a--competitive-advantage-for-indian-msmes/articleshow/104229726.cms?from=mdr

[4] https://www.bizongo.com/solutions/esg-scorecard

[5] https://www.sidbi.in/spex

the intention of MSMEs to adopt or reject pro-sustainability practices. A conceptual model was developed to understand the factors that influence the decision-making behavior of MSME entrepreneurs and business persons.

Conceptual Model

Ajzen (1991) argued that intention precedes adoption of behavior and so any attempt at behavioral engineering requires an understanding of the factors that influence intention. After a careful and exhaustive survey of existing research, the authors developed the following model where intention to adopt sustainable business practices is influenced by the four primary parameters of Performance Benefits, Effort Assessment, Enabling Conditions and Social Pressure. These four parameters are in turn impacted by moderating factors representing the sector the said MSME belongs to, the turnover it accounts for and the geographical location of the MSME. Performance Benefits (PB) are the perceived advantages of green practices covering a large spectrum from lower production costs and energy consumption, improved product quality, better work conditions, stakeholder loyalty, improved access to markets and capital and enhanced competitive edge and energy (Sangwan, 2011; Mittal et al., 2012; Nulkar, 2014). Effort Assessment (EA) indicates the perceived difficulty or ease associated with adoption and implementation of sustainability practices. Short term orientation, Budgetary constraints, lack of trained employees, misunderstanding about circular economy practices, poor investment in research and innovation, and lack of customized technology make it difficult for MSMEs to adopt ESG practices (Lee & Cowling, 2015; Rizos et al., 2016; Environmental Policy Toolkit for SME Greening in EU Eastern Partnership Countries, 2018; Ernst et al., 2022; Vásquez et al., 2024). We explain the conceptual model for sustainable business practices by MSMEs in Fig. 6.1.

Enabling Conditions (EC), like financial incentives,[6] capacity building initiatives,[7] existence of an ecosystem that promotes collaboration between MSMEs and stakeholders positively influence the chances of adoption of ESG, circular economy and other sustainability measures (Panigrahi & Rao, 2018). Basit et al. (2024) identify trained employees, management vision and data driven organisational culture as the main internal enablers and systemic support in the form of government policy, tax incentives and mainstreaming of circular entrepreneurship as the key external enablers that push MSMEs to go green (Hudnurkar et al., 2022). Social Pressure (SP), as a parameter that influences intention of MSME business owners is the force exerted by key stakeholders like environmentally conscious consumers, vendors,

[6] https://www.sidbi.in/head/uploads/spexthegreenpulse_documents/SIDBI-DNBSPeX-Apr-June-2023.pdf

[7] https://msme.gov.in/msmes-are-engines-our-economy-achieve-sustainabledevelopment#:~:text=The%20Ministry%20of%20MSME%20and,Credit%20Linked%20Capital%20Subsidy%20for

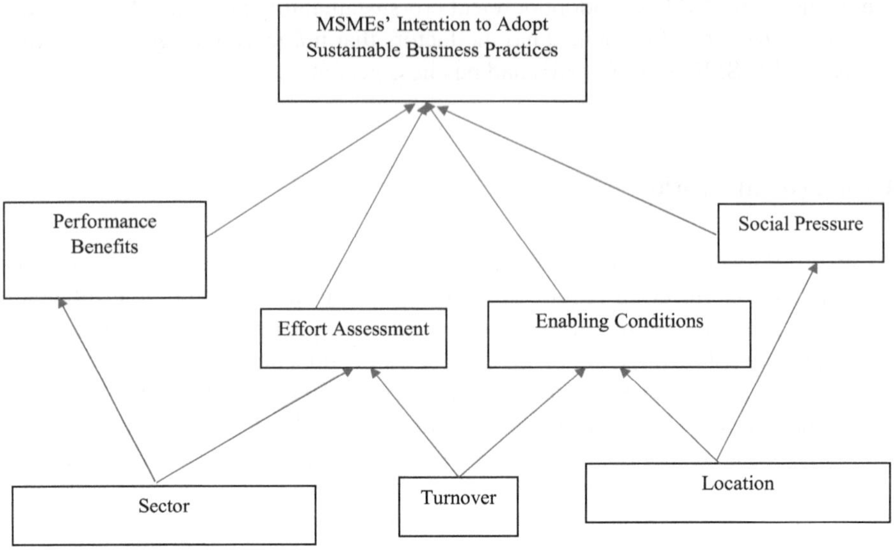

Fig. 6.1 Conceptual Model for Parameters and Moderating factors impacting Intention to Adopt Sustainable Business Practices by MSMEs (own illustration)

media, interest groups and activists to adopt sustainable business practices (Panigrahi & Rao, 2018; Krishnan & Mathiyazhagan, 2022). The conceptual model also includes some moderating factors that impact the core parameters of PB, EA, EC and SP. MSME being a huge umbrella term, the specific sector of the business is likely to impact the perceived benefits and the extent of efforts required to adopt ESG measures (Singh et al., 2015; Hudnurkar et al., 2022). For example, MSMEs in the service sector would find it relatively easier and reap richer benefits of adopting digital technologies, AI and data driven management whereas MSMEs in agri-food spectrum would be more suitable for the adoption of technologies that conserve water (Vásquez et al., 2024). MSMEs with enjoying higher profit margins are more likely to have better access to sustainability aligned technologies and better positioned to implement green business practices, turnover is taken as a moderating factor impacting EA and EC (Malesios et al., 2018). Similarly, 'Location' (Chhikara & Kodan, 2013) representing policies and regulations by local municipalities, state governments, geographic peculiarities and socio-cultural factors, is taken as a moderating factor influencing EC and SP. The model was tested through exhaustive interviews with 41 MSME business owners representing different sectors and regions of India. The respondents in Mumbai were interviewed in person and for those in other regions, video calls were used.

Findings and Discussion

Table 6.1 is the correlation matrix which shows the correlation between the factors with the intention to adopt sustainable business practice. It also shows the inter correlation between the factors. All the factors have a positive correlation with intention to adopt. However, the degree of correlation is higher for performance benefits and effort assessment.

Among the factors, there is a high degree of correlation between Enabling Condition and Effort Assessment, followed by Social Pressure and Performance benefits, and Enabling conditions and Performance benefits. Table 6.1 explains the correlations.

This was corroborated during the in-depth interviews where the respondents strongly emphasised on PB and EA. If the perceived PB is high, like chances of entering a new market or substantive cost reduction, the entrepreneurs were more likely to adopt the sustainable practice. For example, Kishanchand, a farmer with 3 acres of farmland in Kota, Rajasthan, saw opportunity in becoming an agri-preneur when he found the huge difference in the prices organic vegetables attracted. He not only transitioned to organic farming but also set up a successful online business of organic produce. "It was tough to unlearn and relearn farming practices and adopt new methods. But I enrolled for training at Goyal Gramin Vikas Sansthan[8] and became not only an organic farmer but also the owner of a business. I today employ 50 youngsters to manage my farm, inventory, online orders, packaging, delivery and digital promotion", he proudly declared. This case also demonstrates the correlation between EA and EC. Since Kishanchand lived in the vicinity of Goyal Gramin Vikas Sansthan, an organisation evangelising organic agricultural practices, he could benefit from an enabling ecosystem that was more than willing to offer help and mentorship to farmers to transition to sustainable practices. Moreover, the other farmers who had benefitted from the training, provided a huge motivation that helped Kishacnhand to

Table 6.1 Correlation between Intention to adopt and Performance Benefit, Effort Assessment, Enabling Condition, Social Pressure (own data)

	Intention to Adopt	Performance Benefit	Effort Assessment	Enabling Condition	Social Pressure
Intention to Adopt	1				
Performance Benefit	0.6392	1			
Effort Assessment	0.6475	0.4807	1		
Enabling Condition	0.5364	0.5064	0.7950	1	
Social Pressure	0.3601	0.5369	0.3478	0.4124	1

[8] https://ggvsglobal.com/

overcome his apprehension related to effort needed. A similar case was observed for Mansukhlal who was encouraged by his engineer grandson to set up a small plant for manufacturing mycelium packaging solutions using mushrooms. "I was happy running my small grocery business when my grandson, fresh from his engineering college persuaded me to start this business. I reluctantly agreed as I was hestitant to start a new venture in my seventies. However, I was able to see the potential of mycelium packaging emerging as a viable alternative to thermocol, styrofoam and plastic. I am glad I listened to him. The business is good with high margins and I am also contributing towards sustainability", he shared. Here, the growing market for sustainable eco-friendly packaging solutions and technical expertise of his grandson provided the EC which helped him to overcome his hesitancy related to EA. Figure 6.2 shows different levels of performances of MMSEs as explained below.

Figure 6.2 shows the median scores of Performance Benefit (PB), Effort Assessment (EA), Enabling Condition (EC), Social Pressure (SP) across Micro, Small and Medium enterprises. Size of the enterprise and thus the turnover may act as a moderating factors that may influence the impact of PB, EA, EC and SP on intention to adopt. Across all the sizes, PB and SP scores on average are higher. Therefore MSMEs consider PB and SP as relatively more important factors than EA and EC. The PB and SP scores are highest for Small enterprises. The scores of EA and EC are similar across the size of the enterprises, which may indicate that there is not much difference in the average score due to difference in the size of the sector. Micro entrepreneurs are as motivated by high perceived benefits and as vulnerable to social pressure from competition, consumers, vendors and media, as those running small or medium businesses. "I resisted the move to stop using disposable plastic carry bags, for the longest time", confessed Prakash, owner of a restaurant on the outskirts of Hyderabad. "With a steady rush for takeaway orders, plastic bags were the default choice. However, there was a lot of pressure from local NGOs to shun plastic. Now I pack the food in banana leaves and the plastic bags are replaced by cloth bags. Everyone in my area was doing that. I had to fall in line", he explained.

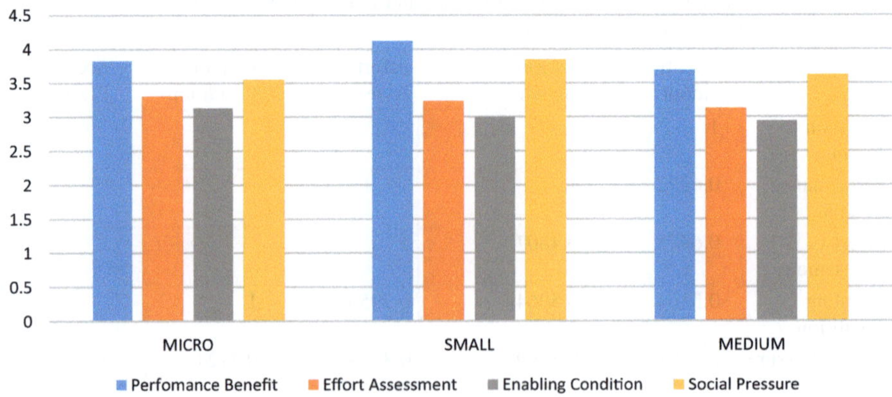

Fig. 6.2 Performance Benefit, Effort Assessment, Enabling Condition, Social Pressure for Micro, Medium and Small Enterprises (Turnover, own illustration)

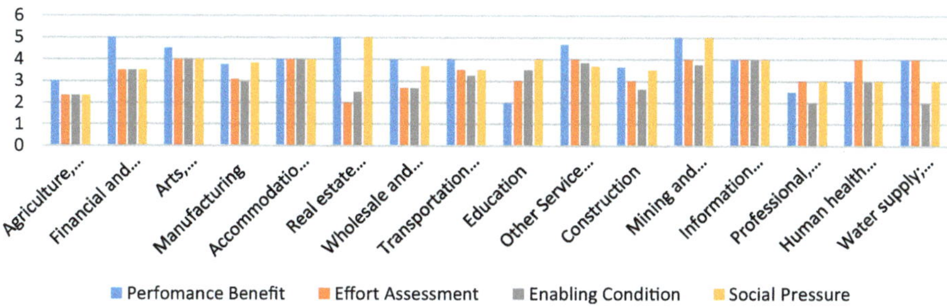

Fig. 6.3 Performance Benefit, Effort Assessment, Enabling Condition, Social Pressure across Type of Industry (Sector, own illustration)

Figure 6.3 shows different levels of performances across type of industry. It shows the median scores of Performance Benefit (PB), Effort Assessment (EA), Enabling Condition (EC), Social Pressure (SP) across by industry type. PB and SP get high scores on average for sectors such as Manufacturing, Real estate activities, Wholesale and retail trade, Transportation and storage, Construction, Mining and quarrying, Arts, entertainment and recreation. For Accommodation and Food service activities and Information and communication, all the factors have equal scores on average. However, for sector like education, EA and EC scores are higher as compared to PB. Overall, we can see a considerable degree of variation of factor scores across industry type, indicating that the sectors perceive this factors differently while considering sustainable business practices.

Figure 6.4 shows different levels of performances across location. The location of the enterprise may also act as a moderating factor, as shown in Fig. 6.4. Again, performance benefits and social pressure on average get high scores for almost all the states. However, EA and EC scores are of comparable magnitudes in states such as Chandigarh and Bihar and higher in Madhya Pradesh. Therefore, the region where the enterprise is located also plays an important role in determining the importance of these factors for intention to adopt ESG practices. Traditionally, entrepreneurship has been concentrated in certain communities in India. Cultural capital and age-old networks make it easier for them to start a new business or diversify/expand existing ones. These, along with policies of local governments could influence the ease or difficulty of doing business in different States. This can be an area of a future study that investigates the differential entrepreneurial ecosystem across India and its impact on adoption of green business practices. In Fig. 6.5 we illustrate responses of entrepreneurs on their apprehensions in incorporating sustainable business practices.

Figure 6.5 shows the reasons that are responsible for the likelihood of the enterprises considering ESG practices. Entrepreneurs have interest in implementing ESG practices but consider them to be expensive and shift focus to other aspect while hampering production. Entrepreneurs also agree that lack of expertise and know-how also act as a deterrent for adopting sustainable practices. Most of the MSMEs exhibit a strong short-term orientation and are more occupied with issues of

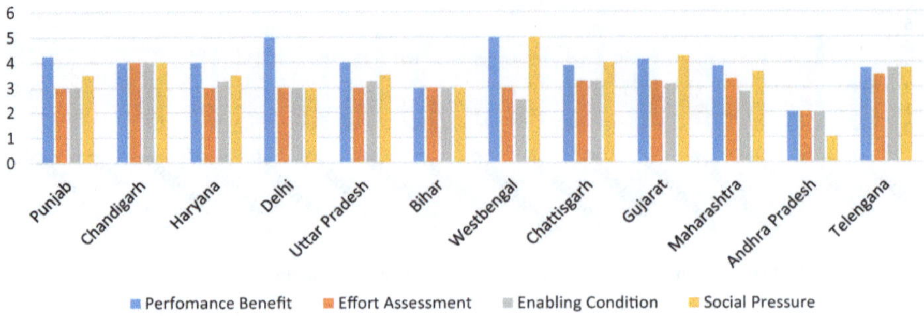

Fig. 6.4 Performance Benefit, Effort Assessment, Enabling Condition, Social Pressure Across Location (States of India, own illustration)

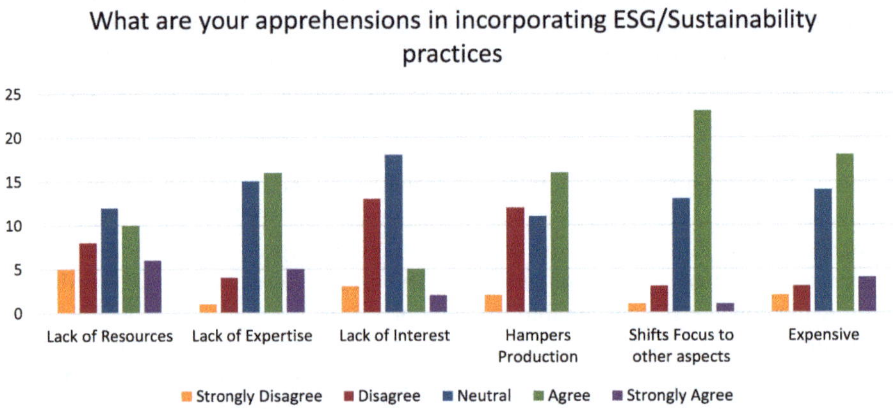

Fig. 6.5 Responses of entrepreneurs on their apprehensions in incorporating sustainable business practices (own illustration)

working capital, negative competition and shrinking markets. For them, sustainable practices are an ideal they hope to achieve in the long run and are not immediately concerned with.

In Table 6.2 we show logit regression results. It presents the results from logit regression model. The binary outcome variable in this model is equal to one if the respondent is willing to adopt sustainability practices in business, equal to zero if not. The logistic regression model estimated is as follows:

$$\ln\left(\frac{p}{1-p}\right) = \beta_0 + \beta_1{}^{*}PE + \beta_2{}^{*}EA + \beta_3{}^{*}EC + \beta_4{}^{*}SP,$$

where p is the probability that a respondent is willing to adopt sustainability practices in business.

For the average median values of PB, EA, EC and SP, the model predicts a probability of about 83% of intention to adopt. The logistic regression model is statistically significant. Out of the four factors, PB and EA significantly influences the

Table 6.2 Logit Regression Estimation and Marginal Impact of Performance Benefit, Effort Assessment, Enabling Condition, Social Pressure on Intention to adopt sustainable business practice (own data)

Logit regression results			
	Coefficient	Standard error	z-statistic
Performance Benefit	1.862	0.823	2.26
Effort Assessment	1.739	0.805	2.16
Enabling Conditions	1.248	1.615	0.77
Social Pressure	0.641	0.934	0.69
Constant	−17.45	7.126	−2.45
Pseudo R square	0.506		
N	41		
LR $chi^2(4)$	25.94		
Prob > chi^2	0.0000		
Marginal Effects			
	dy/dx	Standard Error	z-statistic
Performance Benefit	0.227	0.112	2.04
Effort Assessment	0.212	0.104	2.05
Enabling Conditions	0.152	0.207	0.74
Social Pressure	0.078	0.110	0.71

intention to adopt sustainability practices. A one unit increase in PB, the intention to adopt increases by 22.7 percentage points. For EA, it is by 21.2 percentage points. While the above table helps to understand the relationship between the core four factors of PB, EA, EC, SP and Intention, the graphs preceding it, help to demonstrate how the moderating factors of sector, turnover and location influence the core four factors.

Conclusion

The study offers interesting insights for policy makers and sustainability evangelists to win over more and more MSMEs into their fold and strengthen their intention to adopt green practices. Since the data shows that Performance Benefit is the major driver for ESG adoption, the government can incentivise sustainable business practices through reduced taxes, subsidised transition programs and unambiguous reporting mandates. Information dissemination that reinforces the positives of sustainable practices is needed to educate the MSMEs that sustainability is actually in their favour as it offers longevity to the business and ensures optimum resource management. While pecuniary benefits might be needed in the short term to wean off the MSMEs from their indifference or reluctance to adopt sustainable practices, in the long run, what is needed is a culture shift that promotes environment sensitivity and embeds sustainability in daily living. Another data driven conclusion is the importance of EA. During the interviews, it was shared repeatedly that MSME entrepreneurs tend to get intimidated by the jargon surrounding sustainability. They find the entire discourse overwhelming and disruptive towards their existing business models. Feelings of inadequacy and high perceived difficulties push many MSME owners away from understanding, accepting and adopting sustainable business practices. Instead of alienating the MSMEs, it is pertinent to build an inclusive environment that welcomes them to start their journeys and offers support in the form of training, mentorship and networking. “I will adopt any measure if it is easy”, said Bhushan, an owner of a tannery who was contemplating shutting down his business as he found it increasingly difficult to navigate the sustainability landscape. Ease of adoption is an area that the authors have identified as under invested currently, in need of attention from policy makers, trainers and practitioners. There is an urgent need to demystify sustainability and provide MSME entrepreneurs the ability to identify areas and verticals where it is easy to make a beginning. Digital marketing, energy audits, social responsibility, green supply chain management systems, access to sustainability certifications and public applause for successful transitions can go a long way in positively reinforcing the intention to adopt sustainable practices and the actual translation of intention to behaviour. However, the authors were able to discern an element of apathy and indifference as long as the need for turning their businesses environment friendly remained in the domain of voluntary change. The discussions with the MSME owners pointed towards the need for transparent sustainability policies and reporting standards mandated by the government. A simple but stringent regulatory framework that lays down the roadmap for sector specific guidelines and guarantees ease of adoption and compliance is urgently needed. The authors thus propose a modified conceptual model that incorporates Regulatory Framework as the missing third order moderator that can influence intention and ensure a steady transition to actual adoption. The modified model represented above can be tested in near future as governments across the globe join hands to harmonise the interests of businesses with those of the planet.

References

Abdullah, A., Saraswat, S., & Talib, F. (2023). Barriers and strategies for sustainable manufacturing implementation in SMEs: A hybrid fuzzy AHP-TOPSIS framework. *Sustainable Manufacturing and Service Economics, 2*, 100012. https://doi.org/10.1016/j.smse.2023.100012

Ajzen, I. (1991). The theory of planned behavior. Organizational Behavior and Human Decision Processes, *50*(2), 179–211. https://doi.org/10.1016/0749-5978(91)90020-T

Basit, S. A., Gharleghi, B., Batool, K., Hassan, S. S., Jahanshahi, A. A., & Kliem, M. E. (2024). Review of enablers and barriers of sustainable business practices in SMEs. *Journal of Economy and Technology*. https://doi.org/10.1016/j.ject.2024.03.005

Chhikara, K. S., & Kodan, A. S. (2013). Micro, small and medium enterprises in India: Trends, composition, issues and challenges. *Productivity, 54*(1), 63–76.

Environmental Policy Toolkit for SME Greening in EU Eastern Partnership Countries. (2018). In *OECD green growth studies*. doi:https://doi.org/10.1787/9789264293199-en.

Ernst, R. A., Gerken, M., Hack, A., & Hülsbeck, M. (2022). SMES’ reluctance to embrace corporate sustainability: The effect of stakeholder pressure on self-determination and the role of social proximity. *Journal of Cleaner Production, 335*, 130273. https://doi.org/10.1016/j.jclepro.2021.130273

Hudnurkar, M., Ambekar, S., Bhattacharya, S., & Sheorey, P. A. (2022). Relationship of total quality management with corporate sustainability in the MSME sector: Does innovation capability play a mediating role? *The TQM Journal, 35*(7), 1860–1886. https://doi.org/10.1108/tqm-03-2022-0095

Krishnan, S., Mathiyazhagan, K. (2022). Assessment of Enablers of Additive Manufacturing Implementation. In: Anbanandam, R., Rangnekar, S. (eds) *Flexibility, Innovation, and Sustainable Business. Flexible Systems Management*. Springer, Singapore. https://doi.org/10.1007/978-981-19-1697-7_20

Kumar, S., Raut, R. D., Aktas, E., Narkhede, B. E., & Gedam, V. V. (2022). Barriers to adoption of industry 4.0 and sustainability: A case study with SMEs. *International Journal of Computer Integrated Manufacturing, 36*(5), 657–677. https://doi.org/10.1080/0951192X.2022.2128217

Lee, N., & Cowling, M. (2015). Do rural firms perceive different problems? Geography, sorting, and barriers to growth in UK SMEs. *Environment and Planning C: Politics and Space, 33*(1), 25–42. https://doi.org/10.1068/c12234b

Malesios, C., Skouloudis, A., Dey, P. K., Abdelaziz, F. B., Kantartzis, A., & Evangelinos, K. (2018). Impact of small- and medium-sized enterprises sustainability practices and performance on economic growth from a managerial perspective: Modeling considerations and empirical analysis results. *Business Strategy and the Environment, 27*, 960–972.

Mittal, V. K., Sangwan, K. S., Herrmann, C., Egede, P., & Wulbusch, C. (2012). Drivers and barriers of environmentally conscious manufacturing: A comparative study of Indian and German organizations. In D. Dornfeld & B. Linke (Eds.), *Leveraging technology for a sustainable world* (pp. 97–102). Springer.

Nudurupati, S. S., Budhwar, P., Pappu, R. P., Chowdhury, S., Kondala, M., Chakraborty, A., & Ghosh, S. K. (2022). Transforming sustainability of Indian small and medium-sized enterprises through circular economy adoption. *Journal of Business Research, 149*, 250–269. https://doi.org/10.1016/j.jbusres.2022.05.036

Nulkar, G. (2014). Does environmental sustainability matter to small and medium enterprises? Empirical evidence from India. *International Journal of Environmental Studies, 71*(4), 481–489.

Panigrahi, S. S., & Rao, N. S. (2018). A stakeholders' perspective on barriers to adopt sustainable practices in MSME supply chain. *Research Journal of Textile and Apparel, 22*(1), 59–76. https://doi.org/10.1108/rjta-07-2017-0036

Rizos, V., Behrens, A., van der Gaast, W., Hofman, E., Ioannou, A., Kafyeke, T., Flamos, A., Rinaldi, R., Papadelis, S., Hirschnitz-Garbers, M., & Topi, C. (2016). Implementation of circular economy business models by small and medium-sized enterprises (SMEs): Barriers and enablers. *Sustainability, 8*(11).

Sangwan, K. S. (2011). Quantitative and qualitative benefits of green manufacturing: An empirical study of Indian small and medium enterprises. In J. Hesselbach & C. Herrmann (Eds.), *Glocalized solutions for sustainability in manufacturing* (pp. 371–376). Springer.

Singh, N., Jain, S., & Sharma, P. (2015). Motivations for implementing environmental management practices in Indian industries. *Ecological Economics, 109*, 1–8.

Vásquez, P., Gallego, V., & Soto, J. D. (2024). Transforming MSMEs towards circularity: An attainable challenge with the appropriate technologies and approaches. *Environment Systems and Decisions*. https://doi.org/10.1007/s10669-023-09961-8

Minu Mehta is Dean and Professor at SVKM's NMIMS Anil Surendra Modi School of Commerce, Mumbai, India, one of the most coveted institutions imparting commerce and management education at undergraduate and post graduate levels. An academic leader with more than

three decades of experience, Dr. Mehta holds a PhD in Economics and is a winner of several awards for teaching and research. A leadership trainer and TEDx speaker, she is the recipient of the prestigious AIMS Research and Innovation Fellowship of the Association of Indian Management Schools. Her research interests focus around entrepreneurship, particularly as viewed from the prism of gender, inclusion, and sustainability.

Boishampayan Chatterjee is an economist and holds a doctorate in Economics at Clark University, USA. He holds a postgraduate degree in Economics from Clark University and University of Calcutta, with specializations in Applied Econometrics and Statistics & Econometrics respectively. He is an associate professor at SVKM's NMIMS, Anil Surendra Modi School of Commerce where he teaches economics and quantitative courses.

Chapter 7
Stakeholder Involvement to Foster Sustainable Business Practices in Indonesian MSMEs

Wawan Dhewanto and Rozan Hanifan

Abstract Despite MSMEs' critical contribution to the economy, their integration of sustainable practices is often hindered by challenges such as limited awareness, resources, and incentives. By synthesizing stakeholder theory with theories of the entrepreneurial ecosystem and organizational capability, this research examines the factors that influence MSMEs' sustainability initiatives. This paper investigates the role of stakeholder involvement in enhancing sustainable business practices among MSMEs in Indonesia. Utilizing a mixed-method approach, the study analyzes survey data from MSMEs across West Java, Indonesia, and further validates it using stakeholder interviews. The findings highlight those internal capacities, including digitalization and innovation, are more critical for MSME sustainability than external factors. While external stakeholders are vital for fostering a supportive business environment, internal capacities directly enhance MSME competitiveness and operational efficiency. The paper advocates for prioritizing these internal capacities and calls for reforms in external stakeholder engagement to support MSME growth better. This research contributes to the literature by empirically demonstrating the importance of internal over external factors in advancing MSME sustainability. These insights aim to guide MSMEs and policymakers in promoting the holistic development of the sector.

Keywords Stakeholder involvement · Organizational capabilities · Entrepreneurial ecosystem · Sustainable business practice · Indonesia MSMEs

W. Dhewanto · R. Hanifan (✉)
School of Business and Management, Institut Teknologi Bandung, Bandung, Indonesia
e-mail: w_dhewanto@sbm-itb.ac.id; rozan_hanifan@sbm-itb.ac.id

A. H. Verkuil et al. (eds.), *Core Values and Decision-Making for Sustainable Business*, Sustainable Business Development,
https://doi.org/10.1007/978-3-031-78361-6_7

Introduction

Micro, Small, and Medium Enterprises (MSMEs) play a crucial role in the Indonesian economy. According to the latest data from the Indonesian Ministry of Cooperatives and SMEs (KEMENKOPUKM), MSMEs contribute around 60% to the national Gross Domestic Product (GDP) and absorb more than 97% of the workforce in Indonesia. MSMEs are also the backbone of many local communities, supporting livelihoods and helping to preserve local culture and wisdom. With their significant role, the sustainability of MSMEs is key to inclusive and sustainable economic development in Indonesia (KEMENKOPUKM, 2023).

However, most MSMEs in Indonesia have not yet fully adopted sustainable business practices. Sustainable business practices, which integrate economic, environmental, and social aspects into business operations, are becoming increasingly important to ensure the long-term sustainability of MSMEs and to address challenges such as climate change, natural resource exploitation, and social inequality. MSMEs often face challenges such as a lack of awareness, resources, and incentives to implement more sustainable business practices (Abraham, 2012; Handoko et al., 2017).

Addressing these challenges requires enhanced engagement from a diverse array of stakeholders, including government agencies, financial institutions, non-governmental organizations, business associations, academics, and civil society (Dhewanto et al., 2023a, b). Each of these stakeholders holds the potential to significantly bolster MSMEs in their transition towards sustainable business practices (Noviarto & Samputra, 2021; Yacob et al., 2021). They can provide this support through a variety of initiatives, such as the implementation of supportive policies, the introduction of targeted programs, the provision of financial incentives, and the offering of educational training and mentoring opportunities (Stam & van de Ven, 2021).

This research endeavors to delve into the significance of stakeholder involvement in promoting the adoption of sustainable business practices among MSMEs in Indonesia. It seeks to understand the extent to which both internal and external stakeholders influence these sustainable practices within Indonesian MSMEs. Additionally, the study aims to identify which stakeholders are most crucial in supporting sustainable business practices. Furthermore, it examines how external stakeholders impact the capabilities of internal stakeholders, potentially enhancing their ability to implement and sustain environmentally and socially responsible strategies. By exploring these dimensions, the research will contribute to a broader understanding of the dynamics between stakeholder support and sustainable development within the Indonesian MSME sector.

This research has significance in providing a better understanding of the role of stakeholders in promoting the adoption of sustainable business practices among MSMEs in Indonesia. The research findings are expected to provide valuable insights for policymakers, government institutions, non-governmental organizations, financial institutions, business associations, academics, and other stakeholders in designing effective strategies, programs, and initiatives to support MSMEs in implementing sustainable business practices. Furthermore, this research contributes to the academic literature on environmental management, corporate social responsibility, sustainable development, and MSME development by providing a new perspective from the Indonesian context. The research findings can enrich the understanding of the factors that drive or hinder the adoption of sustainable business practices among MSMEs, as well as the role of stakeholders in this process.

Literature Review

Stakeholder Theory

The stakeholder theory offers a framework for interpreting various dimensions of a firm's strategic behavior, the architecture of management-stakeholder contracts, the institutional mechanisms overseeing these contracts, and the evolutionary dynamics shaping these relationships (Hill & Jones, 1992). Stakeholders are differentiated into internal and external categories, where internal stakeholders are those engaged directly in project execution, and external stakeholders comprise individuals or entities impacted by the project's outcomes (Helbig et al., 2015). They can also be classified based on their power, legitimacy, and the urgency of their claims, distinguishing between definitive, expectant, and latent stakeholders (Phoya & Pietrzyk, 2019).

The critical role of stakeholders in sustainability and their relational dynamics are underscored by Manh et al. (2023), who advocate for a model to investigate stakeholder functions across diverse sectors. In corporate strategy, Oertwig et al. (2017) highlight the evolving responsibilities towards both internal and external stakeholders, emphasizing the necessity for companies to adjust their strategic orientations and operational value creation. Lozano (2018) also stresses the importance of engaging both stakeholder groups within sustainable business models. Additionally, the effective management of corporate identity, which is pivotal for showcasing a company's strengths, is essential for engaging both stakeholder groups (Işildar, 2022).

In the context of this study, the involvement of external stakeholders is examined through the entrepreneurial ecosystem theory, which elucidates the contributions and influences of diverse external agents on the entrepreneurial environment. Concurrently, the role of internal stakeholders is explored through organizational capability theory, which assesses how a company's digitalization and innovation capabilities inherent within an organization bolster its sustainable business practices.

External Stakeholder: Entrepreneurial Ecosystem Theory

Entrepreneurial ecosystem theory has gained significant attention in recent years as a framework for understanding the dynamics that foster high-growth entrepreneurship within regions (Spigel, 2017). This theory focuses on the interconnected set of organizing forces that influence and sustain entrepreneurial activities within a specific geographical area. It is characterized by the collaboration and interaction of various stakeholders, including government entities, private enterprises, investors, consumers, and local communities, to promote economic development, innovation, and entrepreneurship (Mafimisebi & Ogunsade, 2021).

The entrepreneurial ecosystem theory builds upon traditional cluster theory and regional innovation system theory, with a key distinction being its emphasis on the entrepreneur as the central figure (Hammer & Frimanslund, 2022). It considers the ecosystem as a network of interrelated and informally coordinated actors that facilitate and manage entrepreneurial performance within a local environment (Purbasari et al., 2018). The theory also highlights the importance of openness to diverse actors within the ecosystem, as it has been shown that ecosystems characterized by such openness tend to generate more firms (Prokop & Thompson, 2022).

Isenberg's entrepreneurial ecosystem theory, as outlined in various studies (Purbasari et al., 2018; Roundy & Bayer, 2019; Spigel, 2017; Stam & van de Ven, 2021), emphasizes six key domains that are essential for fostering entrepreneurship and innovation within a specific region. These domains include culture, finance, human capital, markets, policy, and supporting infrastructure. The model illustrating these domains is depicted in Fig. 7.1.

Fig. 7.1 Entrepreneurial ecosystem (Isenberg, 2016)

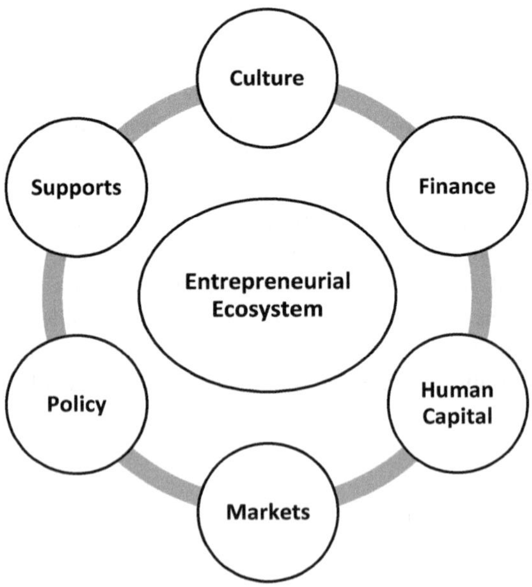

The entrepreneurial culture thrives on values and practices that promote innovation, risk-taking, and a proactive mindset, essential for success in dynamic markets. Entrepreneurial financing involves securing diverse funding sources, such as banks, venture capital, and crowdfunding, crucial for starting, innovating, and expanding businesses. Access to human capital is critical, as it leverages the skills and knowledge of individuals to enhance innovation and competitiveness, shaped by effective recruitment, training, and broader societal influences. Market openness, which ensures the free flow of goods, services, and capital, drives economic growth and global expansion. Entrepreneurial policies support this ecosystem through government regulations and initiatives that stimulate economic development and foster innovation. Finally, supporting business infrastructure provides the necessary resources and facilities that underpin business operations, enabling significant contributions to economic growth (Isenberg, 2016).

Isenberg argues that these elements collectively form the foundation of a conducive environment for entrepreneurial activities to thrive. By focusing on these domains, policymakers, practitioners, and researchers can create entrepreneurial ecosystems that support the growth and success of entrepreneurial ventures. Isenberg's framework underscores the interconnectedness of these domains and how they interact to create a nurturing environment for entrepreneurship (Isenberg, 2016; Spigel, 2017; Stam & van de Ven, 2021).

Internal Stakeholder: Organizational Capability

Organizational capability theory is a fundamental concept in strategic management that focuses on the ability of an organization to effectively utilize its resources to achieve its goals and maintain a competitive advantage over time. This theory draws from various perspectives, such as the resource-based view of the firm, dynamic capabilities, and knowledge-based theories, to understand how organizations can develop and leverage their capabilities for sustained their success (Eisenhardt & Martin, 2000).

Organizational capabilities are crucial for enhancing organizational performance, as indicated by Rehman et al. (2019), who discuss how capabilities mediate the relationship between resources and performance. Andrews et al. (2015) emphasize that organizational capability encapsulates insights from the resource-based theory, underscoring the importance of these capabilities in driving high performance. Moon (2010) stresses the significance of organizational cultural intelligence as a dynamic capability that contributes to sustaining competitive advantages. In this study, organizational capability will be divided only into digitalization and innovation capabilities.

Digitalization Capabilities

Digitalization capabilities encompass the organizational abilities that enable firms to effectively integrate digital assets and business resources, utilizing digital networks to drive innovation in products, services, and processes. These capabilities are crucial for enhancing organizational learning, creating customer value, managing innovation, and ultimately ensuring sustained competitive advantage (Mele et al., 2023). Khin and Ho (2019) emphasize the importance of digital capability in achieving digital innovation, as the success of digital product development heavily relies on a firm's ability to manage digital technologies effectively. This underscores the significance of digital capabilities in driving innovation and competitiveness in the digital era.

In essence, digitalization capabilities draw characteristics from organizational and dynamic capabilities, enabling organizations to sense opportunities, seize them, and reconfigure resources and routines in the context of digital transformation (Annarelli & Palombi, 2021). These capabilities play a crucial role in enhancing organizational adaptability, recognizing, capturing, and reorganizing resources to drive digital innovation and improve overall performance (Sun et al., 2023).

Innovation Capabilities

Innovation capability theory refers to the organizational capacity to manage and integrate various capabilities to foster innovation effectively. It involves the ability to continuously transform new ideas and knowledge into tangible outcomes, such as new products, processes, and systems that benefit the organization and its stakeholders (Bittencourt et al., 2019). This higher-order integration capability allows firms to stimulate innovation by combining key resources and capabilities successfully (Vu, 2020). Innovation capability is intertwined with dynamic organizational capabilities, which consist of organizational routines, processes, tools, and structures that stimulate innovation within the organization (Gullmark, 2021).

Innovation capability is dynamic and contingent on various factors such as market characteristics and innovation novelty (Helge & Breunig, 2017). Understanding customer needs is essential, as innovation capability is closely tied to how well organizations can comprehend and address consumer desires (Marta et al., 2022). Moreover, innovation capability is crucial for gaining a competitive advantage and driving innovation within organizations (Çalık et al., 2017).

Sustainable Business Practice

Sustainable entrepreneurship involves the identification, creation, and utilization of opportunities to develop products and services that not only sustain natural and communal environments but also contribute to the well-being of society (Yulhendri

& Alisha, 2023). It aligns with the Sustainable Development Goals and emphasizes the preservation of nature, life support systems, and communities while pursuing opportunities for economic and non-economic gains (Shepherd & Patzelt, 2011).

Sustainable entrepreneurs aim to transform industries by integrating sustainability into core business activities and striving for mass-market impact beyond niche markets (Hockerts & Wüstenhagen, 2010). Moreover, sustainable entrepreneurship is recognized as a means to address the sustainability challenges of the twenty-first century by balancing economic health, social equity, and environmental resilience through entrepreneurial actions (Strachan, 2018). It involves innovative, market-oriented approaches driven by individuals to create financial and societal value through environmentally or socially beneficial innovations (Schaltegger & Wagner, 2011). This process focuses on solving social and environmental issues to promote sustainable development (Dhewanto et al., 2023a, b).

Model Development

Organizational capability, particularly in terms of digitalization and innovation, is significantly influenced by elements within the entrepreneurial ecosystem in MSMEs. The interactive components of the digital MSME ecosystem are essential for inclusive digital economies, accelerating digitalization in MSMEs (Aminullah et al., 2022). Moreover, the interconnection of green knowledge management and sustainable business capabilities, along with digitalization support, can foster sustainable growth in MSMEs (Alfarizi, 2024). Creating a digital innovation ecosystem enables MSMEs to enhance competitiveness, innovate products, and access global markets.

Entrepreneurial factors positively influence innovation and digital marketing in MSMEs, highlighting the significance of entrepreneurial resilience and innovation capability for sustainable performance (Panjaitan et al., 2022; Usmayanti & Pangestu, 2022). Additionally, market openness positively impacts MSME performance, underscoring the role of these elements in driving organizational success (Anom & Safii, 2022). Furthermore, digital marketing and advanced internetworking programs can help MSMEs increase brand awareness, gain competitive advantages, and expand their market reach (Suryawardani et al., 2021). The synthesis of these references highlights the critical role of elements within the entrepreneurial ecosystem in shaping digitalization and innovation capability in MSMEs. Therefore, the following hypotheses were proposed:

H1a.	*Entrepreneurial culture has a positive significant influence on organizational digitalization capabilities.*
H1b.	*Entrepreneurial culture has a positive significant influence on organizational innovation capabilities.*
H2a.	*Entrepreneurial financing has a positive significant influence on organizational digitalization capabilities.*
H2b.	*Entrepreneurial financing has a positive significant influence on organizational innovation capabilities.*
H3a.	*Access to human capital has a positive significant influence on organizational digitalization capabilities.*
H3b.	*Access to human capital has a positive significant influence on organizational innovation capabilities.*
H4a.	*Market openness has a positive significant influence on organizational digitalization capabilities.*
H4b.	*Market openness has a positive significant influence on organizational innovation capabilities.*
H5a.	*Entrepreneurial policy has a positive significant influence on organizational digitalization capabilities.*
H5b.	*Entrepreneurial policy has a positive significant influence on organizational innovation capabilities.*
H6a.	*Entrepreneurial supporting infrastructure has a positive significant influence on organizational digitalization capabilities.*
H6b.	*Entrepreneurial supporting infrastructure has a positive significant influence on organizational innovation capabilities.*

Moreover, digitalization and innovation capabilities also play a crucial role in influencing sustainable business practices in MSMEs. Previous studies have shown that capabilities such as managerial capability positively impact the financial viability of MSMEs (Ofori-Amanfo et al., 2022). Additionally, the adoption of green knowledge management, green innovation, and digitalization support can contribute to sustainable growth in MSMEs (Alfarizi, 2024). Furthermore, the COVID-19 pandemic has highlighted the importance of eco-innovation capabilities for MSMEs to remain sustainable and competitive (Zulkiffli et al., 2022).

Digitalization factors have been highlighted as influencing the business sustainability of MSMEs (Yaniar et al., 2021). The interplay of strategic green marketing orientation and competitive advantage can impact the engagement with sustainable business practices in MSMEs (Papadas et al., 2019). Furthermore, the adoption of digital financial technologies, such as Fintech, can influence the sustainable business models of MSMEs (Pizzi et al., 2021).

Innovation capability has also been identified as a key factor that mediates the effects of knowledge management, financial literacy, and risk attitude on the performance of these enterprises (Meldona, 2023). Research also emphasizes the significance of innovation in business strategy as a means to create a competitive advantage for MSMEs (Timotius, 2023). Moreover, the role of innovation capability in enhancing the sustainability of MSMEs, especially during challenging times such as the COVID-19 pandemic, has been underscored (Sari et al., 2022). Therefore, the following hypotheses were proposed:

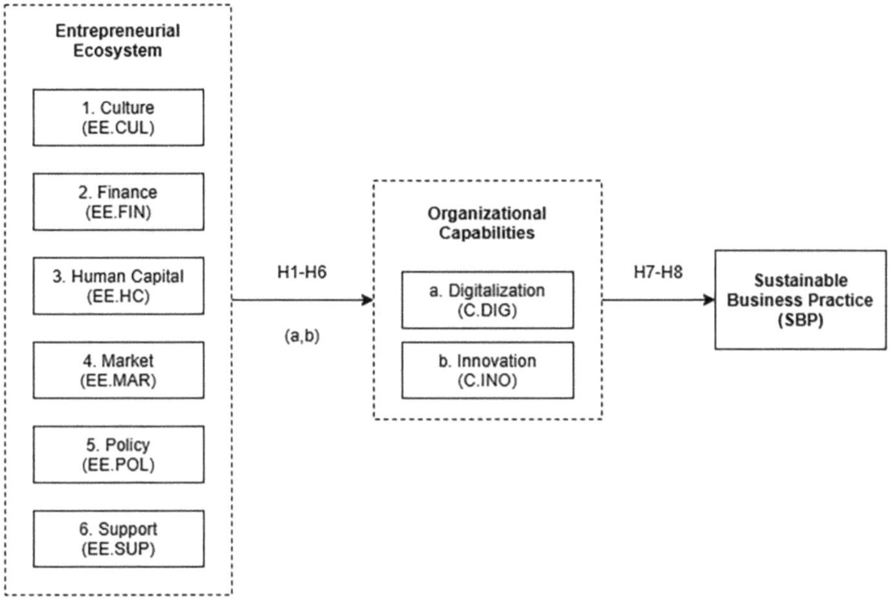

Fig. 7.2 Conceptual Model (own illustration)

H7.	*Organizational digitalization capabilities have a positive significant influence on sustainable business practice.*
H8.	*Organizational innovation capabilities have a positive significant influence on sustainable business practice.*

In summary, this research proposes a conceptual model, as shown in Fig. 7.2.

Methodology

Research Design

This research employs an explanatory mixed-methods approach, beginning with a quantitative phase followed by a qualitative phase to validate the initial findings (Cresswell & Clark, 2018). The study targets MSMEs in West Java, Indonesia, known for its dense concentration of MSMEs, making it an ideal locale for this investigation (BPS Indonesia, 2022). The quantitative phase uses a survey strategy to explore the interrelationships among variables in the study. This method is chosen for its speed, ease, and cost-effectiveness in gathering data across a broad sample of these enterprises (Neuman, 2014).

The study carefully selects its survey participants based on specific criteria, including owners or managers of MSMEs based in West Java with total assets and revenues not exceeding IDR 10 billion and IDR 50 billion, respectively, and employing no more than 100 individuals. The MSMEs targeted are standalone businesses, not franchise branches of larger companies. To collect data, a dual approach of internet-mediated and paper-based questionnaires is utilized through purposive sampling, which ensures the inclusion of respondents likely to provide valuable insights into the research questions. The final sample size for the study includes 326 respondents, allowing for a robust analysis of the strategic behaviors prevalent among MSMEs in the region. Subsequently, qualitative data collection is conducted by selecting several stakeholders within the entrepreneurial ecosystem to validate the findings from the quantitative phase.

Measurement

The study evaluates the elements of the entrepreneurial ecosystem as independent variables, encompassing six areas: Culture (EE.CUL), Finance (EE.FIN), Human Capital (EE.HC), Market (EE.MAR), Policy (EE.POL), and Support (EE.SUP), with measurements adopted from Ahmad & Xavier (2012). Additionally, organizational capabilities such as Digitalization Capabilities (C.DIG) and Innovation Capabilities (C.INO) were assessed following methodologies from Hanifan and Dhewanto (2022) and Pranowo et al. (2020). Lastly, Sustainable Business Practices were measured based on frameworks from Chowdurry (2014) and Kowalska (2020). All measurements utilized a Likert scale ranging from 1 to 5.

Data Analysis Method

For data analysis, this study utilizes the Partial Least Squares (PLS) method using SmartPLS 4.1. The PLS method is particularly well-suited for models that feature complex structures and multiple indicators, as it effectively facilitates the analysis of latent variables. The analysis in this study is conducted in four distinct stages. *First*, the measurement model assessment, which evaluates the reliability and validity of the indicators. *Second*, the structural model assessment examines the relationships between the constructs. *Third*, bootstrapping, statistical technique used to test the hypotheses. Lastly, the Importance-Performance Matrix Analysis (IPMA), prioritizes factors based on their performance and importance to the model's outcomes. Each stage builds upon the previous one to ensure a comprehensive evaluation of the model and its components (Hair et al., 2019). To analyze the qualitative data, a thematic analysis approach is employed, which involves identifying, analyzing, and reporting themes within the data (Saunders et al., 2009). This method provides deeper insights into the quantitative findings, validating and enriching the overall results by uncovering underlying reasons for observed relationships.

Findings and Discussion

Survey Respondent Profile

A significant portion of MSMEs surveyed in this study are relatively young, with 42% aged 1–3 years and 36% aged 3–5 years; only 22% have been operating for over five years. In terms of size, most are small, with 39% employing 1–5 individuals and 49% having 6–19 employees; a smaller fraction (12%) employs 20–99 people. These businesses are predominantly in the Food and Beverage sector, which constitutes 43% of the sample, followed by Fashion (19%), Handicraft (14%), Service (13%), and Retail (10%). Financially, 52% of these MSMEs hold assets ranging from 50 million to 500 million IDR, with 43% reporting revenues between 300 million to 2.5 billion IDR, indicating a diverse yet predominantly small-scale entrepreneurial landscape in the surveyed regions. Table 7.1 illustrates the company profile.

The survey data reveals a slight majority of female participants (54%) compared to males (46%). The respondents are predominantly young, with 49% between 21 and 30 years and 36% aged 31 to 40, indicating a youthful demographic in entrepreneurship. Education levels are high among respondents: 29% hold a Bachelor's degree, 28% have attended college, 33% completed high school, and 9% have a Master's degree. The majority (67%) are business owners, while 33% are managers, highlighting that most respondents hold significant decision-making roles in their organizations. Table 7.2 illustrates the respondents' profile.

Table 7.1 Company profile (own data)

Company profile	Category	Respondents	Frequency
Company age	1–3 years old	137	42%
	3–5 years old	116	36%
	>5 years old	73	22%
Number of employees	1–5 employees	127	39%
	6–19 employees	159	49%
	20–99 employees	40	12%
Field	Food and Beverage	141	43%
	Fashion	62	19%
	Handicraft	46	14%
	Service	43	13%
	Retailer	34	10%
Assets (in IDR)	<50 million	120	37%
	50–500 million	168	52%
	501 million–10 billion	38	12%
Revenue (in IDR)	<300 million	88	27%
	300 million—2,5 billion	140	43%
	2,5 billion–50 billion	98	30%

Table 7.2 Respondents' profile (own data)

Respondent profile	Category	Respondents	Frequency
Gender	Male	151	46%
	Female	175	54%
Age	<20 years old	19	6%
	21–30 years old	159	49%
	31–40 years old	118	36%
	>41 years old	30	9%
Education	High School	109	33%
	College	92	28%
	Bachelor	96	29%
	Master	29	9%
Level	Owner	217	67%
	Manager	109	33%

Measurement Model Assessment

The Measurement Model Assessment was conducted following eight parameters: mean, standard value, outer loading, Variance Inflation Factor (VIF), Cronbach's alpha, composite reliability, Average Variance Extracted (AVE), and Fornell-Larcker Criterion. The results indicate that all indicators performed robustly. Specifically, each indicator reported a mean score above four on a 1 to 5 Likert scale, demonstrating a strong agreement among respondents regarding the variables measured. Furthermore, the standard values for all indicators were maintained below 1.25, ensuring minimal deviation and greater consistency across responses.

Additionally, the assessment revealed that all indicators have outer loadings greater than 0.7, suggesting a high level of individual item reliability within constructs. Also, the VIF values for each indicator were below 5, indicating no multicollinearity concerns that might compromise the validity of the regression results. For the constructs themselves, both Cronbach's alpha and composite reliability scores exceeded 0.7, affirming the internal consistency and reliability of the constructs. Moreover, each variable displayed an AVE higher than 0.5, which confirms adequate convergent validity. These results are systematically tabulated in Table 7.3.

Regarding the Fornell-Larcker Criterion (see Table 7.4), the analysis did not reveal any issues with discriminant validity. The square roots of the AVE of each construct were greater than the construct's highest correlation with any other construct in the model. This suggests that the construct shares more variance with its indicators than with those of any other construct in the model, indicating that the constructs are sufficiently distinct from one another. Table 7.4 shows the Fornell-Larcker Criterion Results.

Table 7.3 Measurement model assessment (own data)

Variable	Indicators	Mean	Std. Dev	Loadings	VIF	CA	CR	AVE
C.DIG	C.DIG1	4.509	0.686	0.811	1.568	0.771	0.853	0.593
	C.DIG2	4.35	0.692	0.737	1.415			
	C.DIG3	4.387	0.69	0.755	1.467			
	C.DIG4	4.42	0.725	0.775	1.53			
C.INO	C.INO1	4.285	0.855	0.788	1.605	0.808	0.874	0.634
	C.INO2	4.261	0.849	0.804	1.689			
	C.INO3	4.307	0.794	0.795	1.646			
	C.INO4	4.147	0.867	0.798	1.628			
EE.CUL	EE.CUL1	4.172	0.897	0.865	2.011	0.817	0.892	0.733
	EE.CUL2	4.322	0.816	0.88	2.137			
	EE.CUL3	4.129	1.046	0.822	1.579			
EE.FIN	EE.FIN1	4.19	1.042	0.845	1.677	0.816	0.891	0.731
	EE.FIN2	4.199	0.997	0.867	1.976			
	EE.FIN3	4.316	0.819	0.852	1.839			
EE.HC	EE.HC1	4.245	0.976	0.884	2.047	0.822	0.894	0.738
	EE.HC2	4.365	0.771	0.85	1.779			
	EE.HC3	4.273	0.887	0.842	1.795			
EE.MAR	EE.MAR1	4.242	0.886	0.871	1.93	0.835	0.901	0.752
	EE.MAR2	4.319	0.797	0.879	2.116			
	EE.MAR3	4.301	0.776	0.851	1.836			
EE.POL	EE.GOV1	4.377	0.728	0.923	3.149	0.906	0.941	0.842
	EE.GOV2	4.267	0.879	0.917	2.86			
	EE.GOV3	4.325	0.955	0.913	2.877			
EE.SUP	EE.SUP1	4.451	0.689	0.84	1.681	0.802	0.883	0.716
	EE.SUP2	4.411	0.741	0.859	1.813			
	EE.SUP3	4.488	0.704	0.839	1.691			
SBP	SBP1	4.528	0.663	0.769	1.578	0.796	0.867	0.619
	SBP2	4.457	0.689	0.806	1.627			
	SBP3	4.534	0.639	0.787	1.664			
	SBP4	4.393	0.722	0.785	1.501			

Structural Model Assessment

The structural model test is implemented to assess the effectiveness of the proposed model. The analysis utilizes R^2 and R^2-adjusted values to measure the model's explanatory power, revealing substantial results with 53.2% for C.DIG, 66.1% for C.INO, and 45.9% for SBP. These percentages illustrate that the model effectively explains the variations in DC, IA, and BR, highlighting its strong explanatory capability. The predictive accuracy and robustness of the model are further affirmed by these findings, which are detailed in Table 7.5.

Table 7.4 Fornell-Larcker criterion result (own data)

		1	2	3	4	5	6	7	8	9
1	C.DIG	0.77								
2	C.INO	0.695	0.796							
3	EE.CUL	0.653	0.726	0.856						
4	EE.FIN	0.617	0.722	0.681	0.855					
5	EE.HC	0.631	0.717	0.761	0.707	0.859				
6	EE.MAR	0.561	0.738	0.741	0.763	0.744	0.867			
7	EE.POL	0.587	0.74	0.796	0.78	0.743	0.821	0.918		
8	EE.SUP	0.657	0.723	0.748	0.73	0.715	0.793	0.798	0.846	
9	SBP	0.625	0.628	0.536	0.477	0.497	0.475	0.45	0.512	0.787

Table 7.5 Structural model result (own data)

Variable	R-square	R-square adjusted
C.DIG	0.532	0.523
C.INO	0.668	0.661
SBP	0.463	0.459

Hypothesis Testing

After confirming the reliability and validity of all measurement items, hypothesis testing was performed using a bootstrapping algorithm with a two-tailed method and 10,000 subsamples. This study includes fourteen hypotheses that delineate the relationships among the nine variables. Each hypothesis was tested at a significance level of 0.05, with some noted at 0.1 for exploratory insights. The outcomes of these hypothesis tests are presented in Table 7.6.

The acceptance of hypotheses H1a and H1b, with substantial path coefficients and strong T-statistics, confirms the significant positive impact of entrepreneurial culture on both digitalization and innovation capabilities. This underscores the pivotal role that a supportive and adaptive culture plays in fostering an environment conducive to adopting new technologies and innovative practices (Aminullah et al., 2022). The statistical strength of these results reinforces the robust influence of entrepreneurial culture as a cornerstone for facilitating technological advancement and creative processes within MSMEs.

Financing continues to be a critical enabler, as demonstrated by the acceptance of H2a and H2b. The positive coefficients indicate that adequate financial resources allow MSMEs to invest in necessary digital tools and innovative methods, aligning with the literature that highlights financing as essential for supporting growth and innovation (Vu, 2020). The significant T-statistics and very low p-values reflect strong empirical support, suggesting that targeted financial interventions can significantly enhance technological and innovative capabilities in MSMEs.

Moreover, human capital also significantly impacts digitalization and innovation, as indicated by the acceptance of H3a and H3b. This suggests that access to quality employees is crucial for effectively implementing and utilizing digital

Table 7.6 Bootstrapping result (own data)

Hypothesis	Path	Path coefficient	T statistics	P values	Decision
H1a	EE.CUL → C.DIG	0.305	3.858	0	Accepted**
H1b	EE.CUL → C.INO	0.189	2.385	0.017	Accepted**
H2a	EE.FIN → C.DIG	0.249	3.21	0.001	Accepted**
H2b	EE.FIN → C.INO	0.206	3.211	0.001	Accepted**
H3a	EE.HC → C.DIG	0.204	2.602	0.009	Accepted**
H3b	EE.HC → C.INO	0.162	2.285	0.022	Accepted**
H4a	EE.MAR → C.DIG	−0.161	1.708	0.088	Rejected
H4b	EE.MAR → C.INO	0.149	1.892	0.059	Accepted*
H5a	EE.POL → C.DIG	−0.144	1.696	0.09	Rejected
H5b	EE.POL → C.INO	0.08	0.867	0.386	Rejected
H6a	EE.SUP → C.DIG	0.343	4.268	0	Accepted**
H6b	EE.SUP → C.INO	0.134	1.814	0.07	Accepted*
H7	C.DIG → SBP	0.364	4.787	0	Accepted**
H8	C.INO → SBP	0.375	4.951	0	Accepted**

*$p < 0.1$, **$p < 0.05$

solutions and for fostering innovation. The results implied the importance of investing in human resources development as a strategy to enhance the innovative and digital maturity of MSMEs.

The mixed results for market openness, with H4a being rejected and H4b accepted at a less stringent significance level ($p < 0.1$), indicate a nuanced impact on digitalization and innovation. While market openness appears to encourage innovation, it does not have a straightforward positive implication for digitalization. This could be due to varying market conditions or external competitive pressures that differently affect these two areas. The findings suggest that while open markets drive innovation by exposing firms to new ideas and competitive challenges, they might not provide sufficient support or incentives for digital transformation.

'The lack of a significant positive impact of the entrepreneurial policy element on digitalization and innovation (both H5a and H5b rejected) might reflect inefficiencies or mismatches between policy intentions and practical implementations. In contrast, the strong positive impact of supporting entrepreneurial infrastructure on both digitalization and innovation (H6a and H6b) underscores the foundational role of robust infrastructure in enabling MSMEs to leverage digital tools and innovate effectively. This distinction between the effects of policy and infrastructure highlights the need for policymakers to reconsider and possibly recalibrate policies to better support technological and innovative endeavors in MSMEs.

Furthermore, the acceptance of H7, with a strong path coefficient of 0.364, demonstrates a robust link between the ability to utilize digital tools and technologies and the enhancement of sustainability within business operations. The p-value of 0 indicates a highly significant result, reinforcing the crucial role of digitalization in driving sustainability. This relationship highlights that digital capabilities enable MSMEs to operate more efficiently, reach wider markets, and implement environmentally friendly practices more effectively, contributing to the overall sustainability of the business.

Similarly, H8 was also accepted with an even higher path coefficient of 0.375, which points to a significant positive influence of innovation on sustainability. This suggests that MSMEs that continuously innovate—not just in terms of products and services but also in their processes and business models—are better positioned to engage in sustainable practices. This finding aligns with broader discussions in the literature, which suggest that innovative firms are more adaptable and can more effectively respond to environmental challenges and market opportunities (Meldona, 2023; Timotius, 2023).

The hypothesis testing results affirm that the entrepreneurial ecosystem's components—culture, financing, human capital, and infrastructure—are integral to fostering digitalization and innovation within MSMEs. These components, when effectively harnessed, can significantly enhance MSMEs' capabilities to engage in sustainable business practices, thereby contributing to their long-term success and resilience. Moreover, both digitalization and innovation capabilities play crucial roles in influencing sustainable business practices among MSMEs. The significant positive effects of both capabilities indicate that they are not only essential for competitiveness and efficiency but are also pivotal in enhancing the sustainability of business operations. This dual impact underscores the intertwined nature of technological advancement and sustainable development, suggesting that MSMEs' efforts to enhance digital and innovative capacities are likely to yield substantial benefits in sustainability.

Importance-Performance Matrix Analysis (IPMA)

The IPMA analysis for this study highlights key areas of business capabilities and the entrepreneurial ecosystem (see Table 7.7). Digitalization and Innovation Capabilities are identified as crucial, both scoring high in importance (0.362 and 0.347, respectively) and performance (85.537 and 85.407), indicating their significant impact on business operations. Entrepreneurial Culture and Supporting

Table 7.7 IPMA result (own data)

Variable	Importance	Performance
C.DIG	0.362	85.537
C.INO	0.347	85.407
EE.CUL	0.137	82.12
EE.FIN	0.122	80.375
EE.HC	0.105	83.112
EE.MAR	−0.002	82.913
EE.POL	−0.013	79.338
EE.SUP	0.129	81.983

Infrastructure also show considerable importance (0.137 and 0.129) but lower performance (82.12 and 81.983), suggesting areas for potential improvement.

On the other hand, Entrepreneurial Financing and Access to Human Capital display moderate importance and performance, indicating these elements are being adequately addressed. Conversely, Market Openness and Entrepreneurial Policy receive low or negative importance scores (−0.002 and −0.013) with performance scores of 82.913 and 79.338, respectively, highlighting these areas as underperforming and less critical, with a particular need for policy improvements to support business practices more effectively. Ultimately, the result of Importance-Performance Matrix is presented in Fig. 7.3, and Table 7.7 shows the results of the Importance-Performance Matrix Analysis.

Qualitative Integration

The qualitative phase of this study includes interviews with six key stakeholders within the entrepreneurial ecosystem: two government officials (G1 and G2), two startup owners (S1 and S2), and two representatives from financial institutions (F1 and F2). These respondents provide a comprehensive view of the ecosystem's dynamics and validate the findings from the quantitative phase.

Government officials emphasized the importance of supportive policies that create a favorable environment for entrepreneurial activities. This indicates that well-structured policies can accelerate digital transformation and encourage innovative practices among businesses. However, the quantitative findings showed that policy

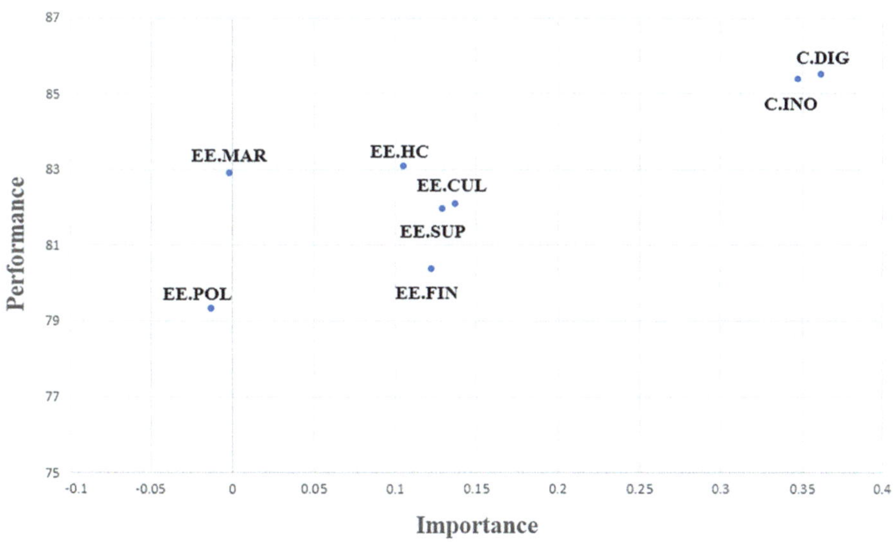

Fig. 7.3 Importance-performance matrix (own data)

elements did not significantly impact digitalization and innovation capabilities, suggesting that gaps between policy formulation and execution may hinder digital transformation efforts. This aligns with findings in the literature that highlight the need for coherent and effectively implemented policies to drive digital transformation (Hellmann & Thiele, 2017).

> Our policies are designed to foster innovation and provide a conducive environment for startups. By streamlining regulations and offering incentives for digital adoption, we aim to create a landscape where MSMEs can thrive and sustain their growth.—G1.

Startup owners highlighted the critical role of financial support in their growth and innovation journeys. Financial resources enable startups to invest in new technologies, facilitating digitalization and innovation efforts. Representatives from financial institutions underscored the significance of a robust financial infrastructure that caters to the unique needs of emerging businesses. A strong financial infrastructure supports startups by providing the necessary capital to explore innovative solutions and adopt digital technologies. These insights are consistent with previous research emphasizing the role of financial support in fostering innovation and technological adoption within SMEs (Clã et al., 2022).

> Access to funding is vital for our growth and innovation. With proper financial backing, we can invest in advanced technologies and sustainable practices that not only improve efficiency but also enhance our competitive edge in the market.—S2.

> We aim to create financial products that meet the needs of emerging businesses. By providing tailored financial solutions, we support MSMEs in their journey towards digitalization and sustainable innovation, ensuring they have the resources to implement environmentally friendly practices.—F1.

A supportive entrepreneurial culture is critical in fostering a mindset that embraces risk-taking and creativity. This culture not only inspires innovation but also encourages the adoption of sustainable methods. This finding supports the view that a strong entrepreneurial culture can significantly enhance innovation and business sustainability (Fellnhofer, 2017).

> A culture that celebrates innovation inspires us to push boundaries. Role models who promote positive business practices help us adopt sustainable methods that are essential for long-term success.—S1.

> Being part of a community that values collaboration and shared learning helps us innovate more effectively. We learn from each other's successes and mistakes, which is crucial for sustainable growth.—S2.

Qualitative findings also find that supporting infrastructure is crucial for operational efficiency and scalability. The presence of robust supporting infrastructure has been linked to higher levels of business innovation and digital adoption (Golubetskaya & Kurlov, 2021).

> Regulations that support digital adoption are crucial for innovation. Our goal is to remove barriers and facilitate a smooth transition to digital platforms, which is essential for the sustainability of MSMEs.—G2.

> Access to reliable utilities and professional services helps us maintain our operations and scale effectively. Without these, innovation would be much more challenging.—S2.

Human capital is another critical element, providing the necessary skills and knowledge to drive technological advancements and innovation. Skilled employees and leaders who understand the potential of digital tools can propel an organization toward greater efficiency and innovation. This underscores the importance of investing in human capital to enhance innovation and sustainable business practices (Capozza & Divella, 2018).

> Human capital is our most valuable asset. Investing in training and development ensures that our team is capable of leveraging new technologies to drive innovation and sustainability.—S1.

> Our commitment to improve MSMEs capabilities in financial and digital literacy ensures that we stay ahead in innovation and provide the best services to our clients. This investment will directly contribute to our sustainability goals.—F2.

Although the quantitative phase revealed that the market element did not significantly impact digitalization capabilities, qualitative insights suggest that market readiness and consumer adoption play indirect roles. While market conditions are crucial for business growth, they do not directly drive digitalization efforts among MSMEs. This finding aligns with the literature indicating that market readiness and consumer adoption are significant for the success of digital transformation initiatives (Jafari-Sadeghi et al., 2021).

> While the market is crucial for business growth, its impact on digitalization is indirect. Many MSMEs still rely on traditional market practices and are slow to adopt digital tools.—S2.

Digitalization and innovation significantly impact sustainable business practices. Adopting digital technologies allows businesses to optimize operations, reduce waste, and improve resource efficiency. Innovation also drives the development of sustainable products and services, fostering long-term environmental and social benefits. These findings align with the literature, which highlights that innovation and technology adoption can enhance efficiency and promote environmentally friendly business operations (Cainelli et al., 2020; Chege & Wang, 2020).

> Digital tools help us track our footprint. By leveraging digital solutions, we can ensure that our business operations are not only efficient but also sustainable.—F2.

> Innovative solutions allow us to address sustainability challenges in unique ways. Our commitment to sustainability is strengthened by the continuous support from stakeholders who encourage eco-friendly practices.—S2.

Overall, the qualitative findings not only validate the quantitative results but also provide deeper insights into how the entrepreneurial ecosystem. The integration of these findings underscores the interconnected nature of the ecosystem elements and their collective impact on fostering a conducive environment for MSMEs to thrive in West Java. This comprehensive view aligns with existing research that emphasizes the importance of a supportive ecosystem in driving innovation and sustainability within SMEs (Aminullah et al., 2022; Panjaitan et al., 2022).

Conclusion and Contribution

This research has delved deeply into the various elements of the entrepreneurial ecosystem and their impact on the digitalization and innovation capabilities of MSMEs. Through rigorous methodology, this study has illuminated the critical factors that influence MSME sustainable practice. The findings underscore the importance of both internal and external stakeholders in shaping business outcomes. However, it is evident from the analysis that internal stakeholder capabilities play a more pivotal role compared to the quality of external stakeholders.

Internal capacities such as digitalization and innovation are crucial for the technological competencies and operational efficiencies of MSMEs. These capacities enable MSMEs to adapt to technological advancements and market changes, significantly enhancing productivity and responsiveness. High importance and performance scores highlight their pivotal role in driving business success. Conversely, external factors, while essential for creating a supportive environment, have a less direct impact. While finance and human resources show moderate performance, market conditions, and policy lag, indicating a mismatch with MSME needs. Entrepreneurial culture and support infrastructure also need improvements to support innovation and adaptability better. The qualitative findings further validate and deepen the understanding of these dynamics.

This research contributes to the theoretical framework by integrating stakeholder theory with entrepreneurial ecosystem and organizational capability theories to analyze the sustainability practices of MSMEs. This study shifts the academic discourse towards a more nuanced understanding of how internal factors are pivotal in driving sustainable practices, challenging previous assumptions that external stakeholder quality predominantly dictates business success (Ferro et al., 2017; Svensson et al., 2016). By demonstrating the direct impact of internal capacities on business sustainability, this research extends existing theories and provides a refined perspective on the dynamics between internal capabilities and external environmental factors in the entrepreneurial ecosystem. This theoretical expansion offers a foundation for future studies to explore the interplay of these dimensions in various business contexts, potentially leading to more tailored strategies for fostering sustainability in MSMEs.

The strategic implications of these findings suggest that MSMEs and policymakers should prioritize strengthening internal capacities to enhance business operations and competitive positioning directly. At the same time, efforts should be made to reform and improve external qualities to create a more supportive and effective ecosystem for MSME growth. These efforts include reforming policies to better align with MSME needs, improving market strategies, and developing robust support infrastructures that are readily accessible to MSMEs.

References

Abraham, R. (2012). Doing business at the base of the pyramid: The reality of emerging markets. *Field Actions Science Report, 4*, 89–96.

Ahmad, S. Z., & Xavier, S. R. (2012). Entrepreneurial environments and growth: Evidence from Malaysia GEM data. *Journal of Chinese Entrepreneurship, 4*(1), 50–69. https://doi.org/10.1108/17561391211200939

Alfarizi, M. (2024). Interconnection of Green Knowledge Management and Sustainable Business Capabilities: An Investigation of the Culinary MSME Sector in the Food Waste Emergency Zone of Indonesia. *IOP Conference Series Earth and Environmental Science, 1324*(1), 12072. https://doi.org/10.1088/1755-1315/1324/1/012072

Aminullah, E., Fizzanty, T., Nawawi, N., Suryanto, J., Pranata, N., Maulana, I., Ariyani, L., Wicaksono, A., Suardi, I., Azis, N. L. L., & Budiatri, A. P. (2022). Interactive components of digital MSMEs ecosystem for inclusive digital economy in Indonesia. *Journal of the Knowledge Economy*. https://doi.org/10.1007/s13132-022-01086-8

Andrews, R., Beynon, M. J., & McDermott, A. M. (2015). Organizational capability in the public sector: A configurational approach. *Journal of Public Administration Research and Theory, 26*(2), 239–258. https://doi.org/10.1093/jopart/muv005

Annarelli, A., & Palombi, G. (2021). Digitalization capabilities for sustainable cyber resilience: A conceptual framework. *Sustainability, 13*(23), 13065. https://doi.org/10.3390/su132313065

Anom, L., & Safii, A. A. (2022). Enhancing MSME performance through market sensing capability, innovation capability, and iconic ethnic product development. *Jurnal Ilmu Manajemen Advantage, 6*(1), 1–10. https://doi.org/10.30741/adv.v6i1.778

Bittencourt, B. A., Zen, A. C., & Prévot, F. (2019). Innovation capability of clusters: Understanding the innovation of geographic business networks. *Review of Business Management, 21*(Special Issue), 647–663. https://doi.org/10.7819/rbgn.v21i4.4016

BPS Indonesia. (2022). Profil Industri Mikro dan Kecil 2020. In *Sensus Pertanian*.

Cainelli, G., D'Amato, A., & Mazzanti, M. (2020). Resource efficient eco-innovations for a circular economy: Evidence from EU firms. *Research Policy, 49*, 103827. https://doi.org/10.1016/j.respol.2019.103827

Çalık, E., Çalışır, F., & Cetinguc, B. (2017). A scale development for innovation capability measurement. *Journal of Advanced Management Science, 5*(2), 69–76. https://doi.org/10.18178/joams.5.2.69-76

Capozza, C., & Divella, M. (2018). Human capital and firms' innovation: Evidence from emerging economies. *Economics of Innovation and New Technology, 28*, 741–757. https://doi.org/10.1080/10438599.2018.1557426

Chege, S. M., & Wang, D. (2020). The influence of technology innovation on SME performance through environmental sustainability practices in Kenya. *Technology in Society, 60*, 101210. https://doi.org/10.1016/j.techsoc.2019.101210

Chowdurry, M. M. H. (2014). *Supply chain sustainability and resilience: The case of apparel industry in Bangladesh*. Curtin University.

Clã, S., Frigerio, M., & Vandone, D. (2022). Financial support to innovation: The role of European development financial institutions. *Research Policy*. https://doi.org/10.1016/j.respol.2022.104566

Cresswell, J. W., & Clark, V. L. P. (2018). *Designing and conducting mixed methods research* (Vol. 31, Issue 4, 3rd ed.). SAGE. https://doi.org/10.1111/j.1753-6405.2007.00096.x

Dhewanto, W., Belgiawan, P. F., Hanifan, R., & Umbara, A. N. (2023a). Strengthening entrepreneurial ecosystem to achieve sustainability through digitalization and innovation: A case of Indonesian MSMEs ecosystem. In *Proceedings of the 3rd Asia pacific international conference on industrial engineering and operations management* (pp. 373–385). https://doi.org/10.46254/ap03.20220063

Dhewanto, W., Hanifan, R., Umbara, A. N., & Zailani, S. (2023b). Sustainable entrepreneurship development strategy for achieving SDGs: Insight from Islamic Boarding Schools

Business Units in Times of Crisis. 2023 IEEE International Conference on Industrial Engineering and Engineering Management (IEEM), 1753–1757. doi:https://doi.org/10.1109/IEEM58616.2023.10406421.

Eisenhardt, K. M., & Martin, J. A. (2000). Dynamic capabilities: What are they. *Strategic Management Journal, 21*, 1105–1120. https://doi.org/10.1108/ebr-03-2018-0060

Fellnhofer, K. (2017). Drivers of innovation success in sustainable businesses. *Journal of Cleaner Production, 167*, 1534–1545. https://doi.org/10.1016/J.JCLEPRO.2017.08.197

Ferro, C., Padín, C., Svensson, G., Varela, J. C. S., Wagner, B., & Høgevold, N. M. (2017). Validating a framework of stakeholders in connection to business sustainability efforts in supply chains. *Journal of Business and Industrial Marketing, 32*(1), 124–137. https://doi.org/10.1108/jbim-12-2015-0253

Golubetskaya, N. P., & Kurlov, A. (2021). Infrastructure support for the innovative transformation of business structures in the digital economy. *Economics and Management, 26*, 1210–1216. https://doi.org/10.35854/1998-1627-2020-11-1210-1216

Gullmark, P. (2021). Do all roads lead to innovativeness? A study of public sector organizations' innovation capabilities. *The American Review of Public Administration, 51*(7), 509–525. https://doi.org/10.1177/02750740211010464

Hair, J. F., Risher, J. J., Sarstedt, M., & Ringle, C. M. (2019). When to use and how to report the results of PLS-SEM. *European Business Review, 31*(1), 2–24. https://doi.org/10.1108/EBR-11-2018-0203

Hammer, S., & Frimanslund, T. (2022). Lessons from a rural ecosystem. *Local Economy the Journal of the Local Economy Policy Unit, 37*(5), 348–363. https://doi.org/10.1177/02690942221147510

Handoko, Y., Herman Soegoto, S., Deden, A., Wahab, S., & Wahdiniwaty, R. (2017). Business sustainable model for MSME in Indonesia. *Journal of Engineering and Applied Sciences, 12*(2), 404–407. https://doi.org/10.3923/jeasci.2017.171.175

Hanifan, R., & Dhewanto, W. (2022). Reinforcing business resilience through entrepreneurial competencies during pandemic COVID-19 : A case of Indonesian MSMEs. *International Journal of Management, Entrepreneurship, Social Science and Humanities, 5*(2), 89–107.

Helbig, N., Dawes, S. S., Dzhusupova, Z., Klievink, B., & Mkude, C. G. (2015). *Stakeholder engagement in policy development: Observations and lessons from international experience* (pp. 177–204). doi:https://doi.org/10.1007/978-3-319-12784-2_9.

Helge, T., & Breunig, K. J. (2017). Conceptualizing innovation capabilities: A contingency perspective. *Journal of Entrepreneurship Management and Innovation, 13*(1), 7–24. https://doi.org/10.7341/20171311

Hellmann, T., & Thiele, V. (2017). *Fostering entrepreneurship: Promoting founding or funding? ERPN: Startup & Small Business Finance (Sub-Topic).* doi:https://doi.org/10.2139/ssrn.2908955.

Hill, C. W. L., & Jones, T. M. (1992). Stakeholder-agency theory. *Journal of Management Studies, 29*(2), 131–154. https://doi.org/10.1111/j.1467-6486.1992.tb00657.x

Hockerts, K., & Wüstenhagen, R. (2010). Greening Goliaths versus Emerging Davids — theorizing about the role of incumbents and new entrants in sustainable entrepreneurship. *Journal of Business Venturing, 25*(5), 481–492. https://doi.org/10.1016/j.jbusvent.2009.07.005

Isenberg, D. J. (2016). Applying the ecosystem metaphor to entrepreneurship: Uses and abuses. *Antitrust Bulletin, 61*(4), 564–573. https://doi.org/10.1177/0003603X16676162

Işildar, P. (2022). Corporate identity structure in hotels: An exploratory study. *International Journal of Contemporary Tourism Research*. https://doi.org/10.30625/ijctr.1118180

Jafari-Sadeghi, V., García-Pérez, A., Candelo, E., & Couturier, J. (2021). Exploring the impact of digital transformation on technology entrepreneurship and technological market expansion: The role of technology readiness, exploration and exploitation. *Journal of Business Research, 124*, 100–111. https://doi.org/10.1016/j.jbusres.2020.11.020

KEMENKOPUKM. (2023). *Profil UKM Indonesia.*

Khin, S., & Ho, T. C. (2019). Digital technology, digital capability, and organizational performance. *International Journal of Innovation Science, 11*(2), 177–195. https://doi.org/10.1108/ijis-08-2018-0083

Kowalska, M. (2020). SME managers' perceptions of sustainable marketing mix in different socioeconomic conditions—A comparative analysis of Sri Lanka and Poland. *Sustainability (Switzerland), 12*.

Lozano, R. (2018). Sustainable business models: Providing a more holistic perspective. *Business Strategy and the Environment, 27*(8), 1159–1166. https://doi.org/10.1002/bse.2059

Mafimisebi, O. P., & Ogunsade, A. I. (2021). Unlocking a continent of opportunity: Entrepreneurship and digital ecosystems for value creation in Africa. *Fiib Business Review, 11*(1), 11–22. https://doi.org/10.1177/23197145211018172

Manh, T. D., Dang, D., Falch, M., Minh, T. T., & Phi, T. V. (2023). The role of stakeholders and their relationships in the sustainability of telecentres. *Digital Policy Regulation and Governance, 25*(2), 104–119. https://doi.org/10.1108/dprg-05-2022-0042

Marta, M. S., Hurriyati, R., & Dirgantari, P. D. (2022). How to understand customer needs? The role of market orientation, innovation capability, and learning orientation. *Relevance Journal of Management and Business, 4*(2), 95–109. https://doi.org/10.22515/relevance.v4i2.4316

Meldona. (2023). Innovation capability and risk attitude mediate the effects of knowledge management and financial literacy on MSME performance. *Journal of Social Economics Research, 10*(4), 194–214. https://doi.org/10.18488/35.v10i4.3555

Mele, G., Capaldo, G., Secundo, G., & Corvello, V. (2023). Revisiting the idea of knowledge-based dynamic capabilities for digital transformation. *Journal of Knowledge Management, 28*(2), 532–563. https://doi.org/10.1108/jkm-02-2023-0121

Moon, T. (2010). Organizational cultural intelligence: Dynamic capability perspective. *Group and Organization Management, 35*(4), 456–493. https://doi.org/10.1177/1059601110378295

Neuman, W. L. (2014). *Social research methods: Quantitative and qualitative approaches* (7th Editio ed.). Pearson Education.

Noviarto, S., & Samputra, P. L. (2021). MSME's sustainable economic behavior for struggling poverty: Agency theory vs. bounded rationality theory. *IOP Conference Series: Earth and Environmental Science, 716*(1). https://doi.org/10.1088/1755-1315/716/1/012120

Oertwig, N., Galeitzke, M., Schmieg, H.-G., Kohl, H., Jochem, R., Orth, R., & Knothe, T. (2017). *Integration of sustainability into the corporate strategy* (pp. 175–200). doi:https://doi.org/10.1007/978-3-319-48514-0_12.

Ofori-Amanfo, J., Akonsi, S. W., & Agyapong, G. K. (2022). The impact of organisational capabilities on the performance of small- and medium-sized enterprises (SMEs). *European Business Review, 34*(5), 642–665. https://doi.org/10.1108/ebr-06-2021-0139

Panjaitan, R., Hasan, M., & Vilkana, R. (2022). Sophisticated technology innovation capability: Entrepreneurial resilience on disaster -resilient MSMEs. *Serbian Journal of Management, 17*(2), 375–388. https://doi.org/10.5937/sjm17-39294

Papadas, K., Avlonitis, G. J., Carrigan, M., & Piha, L. (2019). The interplay of strategic and internal green marketing orientation on competitive advantage. *Journal of Business Research, 104*, 632–643. https://doi.org/10.1016/j.jbusres.2018.07.009

Phoya, S., & Pietrzyk, K. (2019). *Holistic view on multi-stakeholders' influence on health and safety risk management in construction projects in Tanzania*. doi:https://doi.org/10.5772/intechopen.85607.

Pizzi, S., Corbo, L., & Caputo, A. (2021). Fintech and SMEs sustainable business models: reflections and considerations for a circular economy. *Journal of Cleaner Production, 281*, 125217. https://doi.org/10.1016/j.jclepro.2020.125217

Pranowo, A. S., Sutrisno, J., Sulastiono, P., & Siregar, Z. M. E. (2020). The entrepreneurial competency, innovation capability, and business success: The case of footwear industry in Indonesia. *Quality - Access to Success, 21*(178), 20–25.

Prokop, D., & Thompson, P. (2022). Defining networks in entrepreneurial ecosystems: The openness of ecosystems. *Small Business Economics, 61*(2), 517–538. https://doi.org/10.1007/s11187-022-00710-w

Purbasari, R., Wijaya, C., & Rahayu, N. (2018). The impact of the entrepreneurial ecosystem on regional competitive advantage: A network theory perspective. *Russian Journal of Agricultural and Socioeconomic Sciences, 83*(11), 49–63. https://doi.org/10.18551/rjoas.2018-11.07

Rehman, S. U., Mohamed, R., & Ayoup, H. (2019). The mediating role of organizational capabilities between organizational performance and its determinants. *Journal of Global Entrepreneurship Research, 9*(1). https://doi.org/10.1186/s40497-019-0155-5

Roundy, P. T., & Bayer, M. (2019). To bridge or buffer? A resource dependence theory of nascent entrepreneurial ecosystems. *Journal of Entrepreneurship in Emerging Economies, 11*(4), 550–575. https://doi.org/10.1108/jeee-06-2018-0064

Sari, Y., Oktarina, N. Y., Munajat, & Kenamon, M. (2022). The role of innovation capability in MSME sustainability during the Covid-19 pandemic. *International Journal of Social Science and Business, 6*(4), 502–511. https://doi.org/10.23887/ijssb.v6i4.46158

Saunders, M., Lewis, P., & Thornhill, A. (2009). *Research methods for business students* (5th ed.). Pearson Education.

Schaltegger, S., & Wagner, M. (2011). Sustainable entrepreneurship and sustainability innovation: Categories and interactions. *Business Strategy and the Environment, 20*(4), 222–237. https://doi.org/10.1002/bse.682

Shepherd, D. A., & Patzelt, H. (2011). The new field of sustainable entrepreneurship: Studying entrepreneurial action linking "What is to be sustained" with "What is to be developed". *Entrepreneurship Theory and Practice, 35*(1), 137–163. https://doi.org/10.1111/j.1540-6520.2010.00426.x

Spigel, B. (2017). The relational organization of entrepreneurial ecosystems. *Entrepreneurship Theory and Practice, 41*(1), 49–72. https://doi.org/10.1111/etap.12167

Stam, E., & van de Ven, A. (2021). Entrepreneurial ecosystem elements. *Small Business Economics, 56*(2), 809–832. https://doi.org/10.1007/s11187-019-00270-6

Strachan, G. (2018). Can education for sustainable development change entrepreneurship education to deliver a sustainable future? *Discourse and Communication for Sustainable Education, 9*(1), 36–49. https://doi.org/10.2478/dcse-2018-0003

Sun, X., He, Z., & Qian, Y. (2023). Getting organizational adaptability in the context of digital transformation. *Chinese Management Studies, 18*(2), 550–574. https://doi.org/10.1108/cms-06-2022-0222

Suryawardani, B., Wulandari, A., Marcelino, D., Satrya, G. B., Wijaya, R., Lukito, W., & Prabawa, B. (2021). Creative digital marketing and advanced internetworking assistance programs for micro, small, and medium enterprises in Buah Batu District Bandung. *Engagement Jurnal Pengabdian Kepada Masyarakat, 5*(2), 361–376. https://doi.org/10.29062/engagement.v5i2.217

Svensson, G., Høgevold, N. M., Petzer, D. J., Padín, C., Ferro, C., Klopper, H. B., Varela, J. C. S., & Wagner, B. (2016). Framing stakeholder considerations and business sustainability efforts: A construct, its dimensions and items. *Journal of Business and Industrial Marketing, 31*(2), 287–300. https://doi.org/10.1108/jbim-05-2014-0094

Timotius, E. (2023). The role of innovation in business strategy as a competitive advantage: Evidence from Indonesian MSMEs. *Problems and Perspectives in Management, 21*(1), 92–106. https://doi.org/10.21511/ppm.21(1).2023.09

Usmayanti, V., & Pangestu, M. G. (2022). MSMEs digital marketing and entrepreneurship's factor of performance: Jambi (Indonesia) perspective. *Jurnal Ekonomi Pendidikan Dan Kewirausahaan, 10*(2), 155–166. https://doi.org/10.26740/jepk.v10n2.p155-166

Vu, H. M. (2020). A review of dynamic capabilities, innovation capabilities, entrepreneurial capabilities and their consequences. *Journal of Asian Finance Economics and Business, 7*(8), 485–494. https://doi.org/10.13106/jafeb.2020.vol7.no8.485

Yacob, S., Sulistiyo, U., Erida, E., & Siregar, A. P. (2021). The importance of E-commerce adoption and entrepreneurship orientation for sustainable micro, small, and medium enterprises in Indonesia. *Development Studies Research, 8*(1), 244–252. https://doi.org/10.1080/21665095.2021.1976657

Yaniar, N., Sasono, A. D., Sukoco, A., & Rosyid, A. (2021). Financial technology and digital marketing on MSMEs and their impact on financial performance and business sustainability. *Ijebd (International Journal of Entrepreneurship and Business Development), 4*(6), 828–838. https://doi.org/10.29138/ijebd.v4i6.1546

Yulhendri, Y., & Alisha, W. P. (2023). Sustainable entrepreneurial ecosystem: Systematic literature. *Review, 340–362*. https://doi.org/10.2991/978-94-6463-158-6_31

Zulkiffli, S. N., Zaidi, N. F. Z., Padlee, S. F., & Sukri, N. K. A. (2022). Eco-innovation capabilities and sustainable business performance during the COVID-19 pandemic. *Sustainability, 14*(13), 7525. https://doi.org/10.3390/su14137525

Wawan Dhewanto, Prof. PhD., earned his bachelor's degree in Industrial Engineering from Institut Teknologi Bandung (ITB), Indonesia. He continued his studies at TU Delft, Netherlands, on a Huygens scholarship, earning a MSc in 2003, and later completed a PhD in 2012 at Monash University, Australia, on an Australian Development Scholarship. He joined ITB's School of Business and Management (SBM) in 2004, leading initiatives in Entrepreneurship and Technology Management research group. He became Professor of Entrepreneurship in 2021, with research interests in Entrepreneurship, Start-up Development, SME Scale-up, Entrepreneurial Leadership and Business Innovation.

Rozan Hanifan, MSM, is a dedicated professional in Entrepreneurship Studies, making significant contributions to research in innovation. He has held key roles as a Researcher at SBM ITB, Agile Innovation Labs, and currently working as a Business Development Manager at Ganesa Research and Consulting. Rozan has been integral to the PT Pelita Air Services Spin-off Project and Curriculum Development for Entrepreneurship Developer at the Ministry of Cooperative and Small Medium Enterprise (KEMENKOPUKM). Recognized for his excellence, he was one of the fastest graduates in the Master of Science in Management (MSM), ITB Indonesia 2021 cohort and received the JASSO Scholarship for Global Project-Based Learning in Japan. His published works in international journals and conference presentations underscore his commitment to advancing Entrepreneurship Studies.

Chapter 8
Barriers to Business Model Innovation: Insights from SMEs in Switzerland

Karsten Eichler, Dario Meyer, and Rolf Meyer

Abstract Business model innovation (BMI) has become increasingly important in recent years, but the question remains as to why only a small number of companies are doing it. In addition, the rate of companies innovating in general is decreasing in Switzerland, especially among small and medium-sized enterprises (SME). Therefore, the aim of this study was to investigate the barriers to business model innovation among small and medium-sized enterprises in Switzerland. A quantitative approach with a written survey sent to 4000 companies was adopted. A sample of 405 company managers from the Swiss canton of Schwyz responded. The data were statistically analysed using SPSS.

The results show that medium-sized companies and those in high-tech sectors are more likely to engage in business model innovation. In contrast, smaller companies and those in less technical sectors face more challenges. The analysis identified seven barriers to BMI that are significant. Two barriers turned out to be triggers for BMI rather than actual obstacles. Past barriers were more intrinsic, related to diversity, risk aversion and own expertise. For current BMI plans, the most relevant barriers are more likely to be related to cost, technology and capacity. Additionally, the relevance of BMI may increase as companies focus on data and digital business models. These findings can assist companies in identifying and overcoming barriers in the innovation process. Furthermore, organisations such as economic development agencies or higher education institutions can adapt their services to better support SMEs in overcoming these barriers and fostering innovation.

Keywords Business model innovation · Innovation barriers · Small and medium sized companies · Quantitative research - Switzerland

K. Eichler · D. Meyer (✉) · R. Meyer
University of Applied Sciences and Arts Northwestern Switzerland (FHNW), Windisch, Switzerland
e-mail: author.ke@bluewin.ch; dario.meyer@fhnw.ch; rolf.meyer@fhnw.ch

A. H. Verkuil et al. (eds.), *Core Values and Decision-Making for Sustainable Business*, Sustainable Business Development,
https://doi.org/10.1007/978-3-031-78361-6_8

Introduction

Entrepreneurial success is an important factor in sustainable economic development, which ultimately benefits society as a whole. In the long term, it promotes economic stability, creates jobs and fosters innovation, all of which contribute to a thriving economy. When businesses are successful, they are more likely to invest in sustainable practices, provide support for community development and drive social progress. Therefore, a company's ability to be entrepreneurial and innovate supports the broader objective of sustainable development and societal well-being (Acs et al., 2013; Huđek & Hojnik, 2020; Porter & Kramer, 2011).

In this context, innovating business models (BMI) is also crucial for the long-term success of companies in a constantly changing economic environment (Bereznoi, 2015). In an era of rapid technological advancement and increasing market dynamics, companies must continuously adapt and innovate their business models in order to remain competitive (Bashir & Verma, 2017). Technological developments such as digitalisation, artificial intelligence and the Internet of Things are constantly opening up new opportunities and challenges that require a flexible and innovative approach to business models. This is not just about optimising existing models, but also about finding completely new ways to create and deliver value.

Despite the obvious need for innovation, many companies face significant barriers to implementing new business models. These barriers can be very diverse and include, for example, organisational, institutional and technological challenges as well as external market conditions. Organisational barriers can include resistance to change, inadequate organisational learning mechanisms and lack of innovation capabilities (Rüb et al., 2017). Institutional barriers often concern the willingness to take risks and the culture of innovation within the company (Bocken & Geradts, 2020). Technological barriers can include difficulties in integrating new technologies and adapting existing systems (Mosig et al., 2021). In addition to these internal challenges, external market conditions also play a role. External barriers include market conditions, regulatory restrictions or supply chain issues (Guldmann & Huulgaard, 2020).

In Switzerland, the number of companies engaged in innovation has fallen significantly over the past 20 years. This is particularly true for SMEs, where the number has halved (SBFI, 2020). Furthermore, BMI is only carried out by a minority of companies (Meyer & Meyer, 2020). A previous study found that knowledge of BMI is insufficient and thus acts as one possible barrier, but left open the question of why 56% of companies that claim to have no difficulty in innovating have not innovated in the last 10 years and why the number of innovative companies is declining sharply (Meyer et al., 2023). By focusing on one Swiss canton, this study aims to identify and analyse the current barriers to business model innovation, with an additional focus on big data as part of digitalisation and a possible enabler of innovation (SBFI, 2020). The findings will help companies to overcome these challenges and successfully implement innovative business models. The following research questions will be answered:

- Q1: What is the current status of the planned BMI projects?
- Q2: Why was there no investment in more business model innovation in the past?
- Q3: Are today's barriers different from those of the past?
- Q4: Does the use of big data, or lack thereof, have an impact on business model development?

This chapter is divided into five more sections. Firstly, the known barriers to BMI are identified and a theoretical framework developed that was used for the survey. Section "Methods" described the methodology in collecting and analysing the data. Section "Results" presents the results and defines the barriers to BMI. The final two sections discuss the results considering the current literature and conclude the chapter with further research directions and limitations.

Literature Review

Business Model Innovation

With regard to the multidimensionality of BMI, there are different definitions in the research landscape, with a lack of consensus (Bashir et al., 2020). According to Teece (2010), business model innovation (BMI) entails rethinking and redesigning how a business creates, delivers, and captures value. This can involve launching innovative products or services, tapping into new markets, or using technology to enhance existing business processes. Similarly, Euchner and Ganguly (2014) highlight that BMI is a strategic redesign of a firm's value proposition, the way it creates value, and how it captures value in the market. Business model innovation can be incremental, involving minor adjustments, or radical, involving a significant overhaul of a company's business structure (Souto, 2015). Incremental innovation focuses on improving existing processes, while radical innovation often leads to entirely new ways of conducting business (Souto, 2015).

A common differentiation in the definition is between single and multi-dimensional changes. For example, one definition focuses on Osterwalder's business model canvas and argues that a BMI already happens when a company innovates in at least one of the nine aspects; thus, multi-dimensionality is not taken into account (Rüb et al., 2017). For this paper we define a business model in terms of four dimensions: customers, value proposition, value chain and profit mechanism. A BMI, according to this definition, changes at least two of these four dimensions and thus allows the company to define new rules in the market (Philippi & Hinz, 2018). This allows a clear differentiation from other types of innovation, such as product or process innovation. Thus, the definition that was also given to the survey participants was the following:

> A business model innovation changes at least two of the four dimensions: customer base, value proposition, value chain and revenue model. This allows the company to define new rules in the market. Key aspects are, for example, the novelty of the business model, the

> barriers to change for the customer, or the efficiency of the company's own processes or organisation. The advantage of business model innovation is that the long-term positive effect for the company is greater than with individual product or process innovations.

Bashir et al. (2020) identified five key aspects around business model innovation: triggers, barriers, enablers, dimensions and outcomes. This paper focuses solely on the aspect of the barriers, meaning the factors that hinder a business model innovation at all possible stages.

The use of big data is becoming increasingly important for companies, especially due to the increasing availability of a wide variety of data and its significant impact on the value chain (Matzler et al., 2018). Big data refers to very large and complex datasets that are difficult to process and analyse using traditional data processing tools and techniques. These datasets can come from various sources, such as company specific devices and sensors, social media, geolocation data or transactions (Chaudhary et al., 2015). Big data has started in IT, but is now used more for business transformation, where its application has helped businesses to succeed (Chaudhary et al., 2015). There is an opportunity to design BMI around the collection, analysing and interpretation of external as well as internal data (Sorescu, 2017). The question of associated BMIs therefore also arises for small and medium-sized enterprises, especially as BMI itself is also becoming increasingly important across sectors (Philippi et al., 2022) and, in the context of big data, is also related to the aforementioned Q4.

Barriers to Business Model Innovation

Research has shown different barriers to business model innovation. Chesbrough (2010) in his influential work states that the biggest barrier to BMI is resistance to change within the organisations. The Swiss State Secretariat for Education, Research and Innovation identifies the three main barriers to innovation as "*high costs*, *lack of own funds* and *shortage of skilled workers*" (Spescha & Wörter, 2020). Also, Morabito (2015) and Friedrich von den Eichen et al. (2015), in partial agreement with the SBFI report and Chesbrough, identify the biggest BMI innovation barriers as lack of awareness of BMI, entrenched thinking (need not seen), bureaucratic hurdles in internal processes, lack of consideration of the customer perspective, lack of expertise, cultural differences, organisational structure and technological maturity. As these examples show, there are different hurdles concerning various areas of a company.

Bashir et al. (2020) have identified six categories of barriers in their systematic literature review. Table 8.1 shows the categories and the possible barriers identified in research for each category.

From these barriers, the theoretical framework at the end of this section has been developed.

Table 8.1 Barriers to business model innovation structured into six categories (Bashir et al., 2020; Chesbrough, 2010; Foss & Saebi, 2016; Lopez et al., 2019; Morabito, 2015; Spescha & Wörter, 2020; Friedrich von den Eichen et al., 2015)

Category	Barriers
Cognitive	• Lack of know-how and awareness • The need is not seen • Complexity of business models • Risks are higher than the cost
Organizational	• Organizational resistance • Bureaucratic hurdles of internal processes • Lack of expertise • Fully occupied by day-to-day-business
Institutional	• Business strategy • Lack of funds • High cost of innovation • Shortage of skilled workers • Cultural differences
Market	• Lack of information • Monopolies • Relative cost of labor • Too complex customer base • Customers are not open to new developments • Market does not allow adjustments
Behavioral	• Lack of perceived control • Lack of attention
Technological	• Technical know-how • Cost of technological solutions

Context Business Landscape Canton Schwyz

The survey was conducted in the Swiss canton of Schwyz. Schwyz is a small canton with slightly above 160′000 inhabitants. 5% work in the primary economic sector, 25% in the secondary and 70% in the third sector, offering services (Federal Statistical Office, 2024). The canton has 20′600 registered companies, the majority of which are micro companies with less than 10 employees (Kanton Schwyz, 2023). Considering the nature of the company sample for the canton of Schwyz, the further reflections and the empirical analysis are focused on SME's and services.

Heikkilä introduces three types of SMEs (Heikkilä et al., 2017) with different focus on specific BMI activities, respectively business model dimensions according to Philippi & Hinz (2018). Firstly, there are the profitability seekers (Heikkilä et al., 2017), who focus on value chain and yield mechanics. They begin with reducing costs and enhancing the effectiveness and efficiency of their main operations and assets. Following this, they attempt to optimize their pricing strategies. Eventually, the attention shifts from internal improvements to a stronger customer orientation, aiming to update their offerings. This shift may result in changes to the value chain. Secondly, the growth seekers have the customers and the value proposition at their core (Heikkilä et al., 2017). They initially concentrate on their existing customers in the existing markets. Subsequently, they work on enhancing their offerings in the current markets or on exploring other markets including going international. The subsequent steps involve identifying potential partners and the appropriate channels to serve these additional customers. Their primary focus remains on the customer side. And the third category are new business starters that look at all four dimensions of the business model, namely customers, value proposition, value chain and yield mechanics (Heikkilä et al., 2017). They cyclically build their business model, covering most components by analysing and testing its viability. They focus on improving the entire business model rather than just a few components. In the early stage, they rely on a few first movers and maintain agility to serve them effectively.

As indicated in the report published by the Swiss State Secretariat for Education, Research and Innovation (SBFI, 2020), the service industry is characterised by significant heterogeneity, which presents a challenge in identifying shared characteristics of innovation activities within this sector. Nevertheless, the following four characteristics appear to be common to a significant proportion of service businesses:

- The subordinate role of research and development as a source of innovation. In many service industries, the development of new practical solutions is not a primary objective; rather, the focus is on providing new instruments to offer services to customers.
- For the vast majority of service providers, the use of information and communication technology is important. It now plays a much more important role within the service sectors. The realisation of opportunities associated with it depends, however, on several factors that determine success. Nevertheless, the development and adaptation of some intangibles (such as human capital, organisational structures, and business models) is crucial to fully exploit the productivity potential of information and communication technology.
- Innovation in services largely involves new business models and is mostly focused on changing organisations and processes.
- The demand side and thus the users of the innovations play an increasingly important role in their development and implementation.

In addition to the similarities a very important difference exists comparing traditional and modern services. The latter require to a certain extent the building of expert knowledge that must be maintained and updated in order to have a competitive advantage. Consequently, modern services can be classified as so-called Knowledge-Intensive-Services (KIS).

Knowledge has become a decisive competitive factor for modern services (SBFI, 2020). Successful companies no longer primarily produce and sell physical products, but respond to their customers' needs and solve their problems with highly functional and intelligent products. In this process, knowledge becomes the most important production factor. KIS have specific characteristics that distinguish them from other services: they use cumulative learning processes with their customers, e.g. co-creation or co-invention. The solutions developed in this way are tailor-made and cannot be easily reproduced. An important prerequisite for success in this process is mutual trust between the service provider and the customer. The exchange between them gives the service provider access to sensitive, business-relevant knowledge of the customer (SBFI, 2020). Using this knowledge, the literature indicates that SMEs can enhance their performance by engaging in BMI (Andersen et al., 2022). However, SMEs often fail to realise the impact due to limited resources, financially and personnel-wise, which limit their abilities.

Theoretical Framework

In summary, it is to be expected that a certain reluctance towards BMI will be observed, due to the size of the companies represented in the sample. This may also have an impact on the answers provided to questions Q1 and Q2. The specific obstacles to innovation may vary depending on the selected route (profitability, growth, or new business) and the degree of emphasis placed on knowledge (KIS) and data affinity. These factors could influence the responses to questions 3 and 4. The barriers identified in Table 8.1 were adapted into a theoretical framework for use in a questionnaire. Also, the SME type, KIS focus and data affinity were added to the framework.

Table 8.2 illustrates the theoretical framework with the anticipated linkage of innovation barrier to SME Type, KIS Focus and Data Affinity.

The assessments of SME type, KIS focus and data affinity were made by the authors of this paper based on their descriptions as explained in the literature above.

Table 8.2 Theoretical framework: anticipated linkage of innovation barrier to SME Type, KIS Focus and Data Affinity (own illustration)

Linking barriers to SME type, KIS focus and data affinity			
Barrier	SME type	KIS focus	Data affinity
We see no need for an adjustment.	No BMI interest	No	No
Our employees are sceptical about new things.	Profitability	No	No
Our teams are too culturally diverse.	Profitability	Yes	Yes
We invest cautiously because of the high costs.	Profitability	Yes	Yes
We do not have the own funds.	All	Yes	No
The risks do not justify the expected benefits.	Growth	No	Yes
We lack the expertise in implementation.	New Business	Yes	Yes
We do not have the necessary technology.	New Business	Yes	Yes
We are fully occupied with the day-to-day business.	All	No	No
Our internal processes are complex.	Profitability	Yes	No
Our internal structure is complex in design.	New Business	No	Yes
Our customer base has a complex structure.	Growth	Yes	No
Our customers are sceptical about new things.	Growth	No	No
The market does not allow any adjustment.	Growth	No	No

Methods

Sample and Data

The survey sample initially comprised 4000 companies, spanning various sizes and industries in the canton of Schwyz. The dataset was obtained from the Swiss Federal Statistical Office, which provided a random sample from the company register. Due to ongoing liquidations, the number of addressed companies was reduced to 3947. Additionally, 104 companies were excluded due to undeliverable mailings, resulting in a final count of 3843 contacted businesses. Table 8.3 shows survey sample data.

The survey received 405 analysable responses, which equals to a response rate of 10.54%. In terms of response methods, the survey used both online and postal channels. Of the 405 usable responses, 133 were submitted online and 272 were received by traditional mail. In the following sections of this paper, if the number of responses (N) is less than 405, it means that a subset of respondents did not respond to that particular question. However, they did respond to the other questions, so that the overall feedback can be analysed with a 95% confidence level and a 5% margin of error (in this case, the minimum is 350).

Measures of Variables

The survey was developed based on the literature and the framework developed in Table 8.1. The data were measured using a mix of questions. On the one hand there were simple yes/no, single and multiple choice questions. On the other hand, there were some Likert scale questions with a choice on a scale from agree to disagree. At the beginning, participants were asked if they knew what BMI was. This was

Table 8.3 Survey sample data (own data)

Survey sample	
Companies in the sample	4000
Ongoing liquidations	−53
Addressed companies	3947
Undeliverable mailings	−104
Contacted companies	**3843**
Analysable responses	**405**
Response rate **(Analysable responses/contacted companies' ratio)**	**10.54%**
Online responses	133 (32.84%)
Mail responses	272 (67.16%)

followed by a definition of BMI to ensure that all participants answered the subsequent questions with the same knowledge. The same approach was used for the application of big data. Participants were given a definition before being asked if they use big data.

Statisical Analysis

Participants in the quantitative survey could respond online via QR code or URL or by returning a paper questionnaire using a provided envelope. The responses from both channels were combined into a single raw data file for further analysis.

The following statistical methods were applied to analyse the collected data using IBM SPSS Statistics: bivariate logistic regression, crosstabulation (Pearson's Chi-Square) as well as rank correlation (Spearman's Rho) (Field, 2017). To ensure the correct application of these methods, several variables underwent data transformations to meet the requirements of the selected statistical techniques.

Company Categories

To get meaningful results two classifications were made in the preparation of the survey. On the one hand, the company sizes were classified according to the definition of the European Commission, without looking at financial data, focusing only on the number of employees (European Commission, 2023). Table 8.4 shows the company size classification.

On the other hand, the industries were grouped following the definitions of the Swiss State Secretariat for Education, Research and Innovation (SBFI, 2020):

Low-tech industry:

Food, beverages and tobacco/Textiles, clothing and leather/Wood, paper and printed products/Coke and refined petroleum products/Glass and glass products, ceramics, processing of stones and earths/Metal production/Metal products/Equipment, articles for the installation and repair of machinery.

High-tech industry:

Chemicals, pharmaceuticals/Rubber and plastic products/Data processing equipment, electronic and optical products (including watches and electromedical

Table 8.4 Company size classification (own data)

Company size by number of employees	
Micro	1–9
Small	10–49
Medium-sized	50–249
Big	250 and more

equipment)/Electrical equipment/Mechanical engineering vehicles (including motor vehicles).

Traditional services:

Wholesale and retail trade, sale, and repair of motor vehicles/Transport (freight and passenger) and warehousing/Postal services/Hotels and restaurants/Craft/ Cleaning.

Modern services:

Publishing (including software publishing), production and broadcasting of audio-visual content, telecommunication content, telecommunications, information services (including programming and consulting, accommodation and data analysis)/Financial, insurance and reinsurance services (including pension funds)/ Professional services in the profession of law, accounting and consulting/Rental and leasing services (including vehicles and private accommodation)/Booking services/Marketing services/Administrative support services/Architectural and engineering activities/Scientific and R&D activities/Therapeutic services.

Energy:

Electricity, gas, water supply, sewerage, and waste management services.

Construction:

All activities in building construction, civil engineering, and specialised construction activities.

Results

The following chapter shows the results of the survey.

Demographics and Outlook

Two thirds (77%) of the responding business owners own micro enterprises with less than 10 employees (including one-person enterprises), as shown in Fig. 8.1. The remaining third are between small and medium-sized enterprises. In summary, most enterprises surveyed are micro, while medium-sized enterprises with more than 49 employees are the least common in the sample. Companies with more than 250 employees were excluded from the sample. Figure 8.2 shows the distribution of various industry sectors among the sample. The largest sector is modern services (33%), followed by traditional services (29%) and construction (23%), while energy is the least represented sector in the sample. This means that approximately 85% of the enterprises in the sample provide services to their clients if construction is included. Figure 8.1 illustrates the company size distribution in the sample, and Fig. 8.2 shows the industry distribution in the sample. Figure 8.3 illustrates the turnover trend in the sample of companies, and finally, Fig. 8.4 shows the profit trend in the sample of companies.

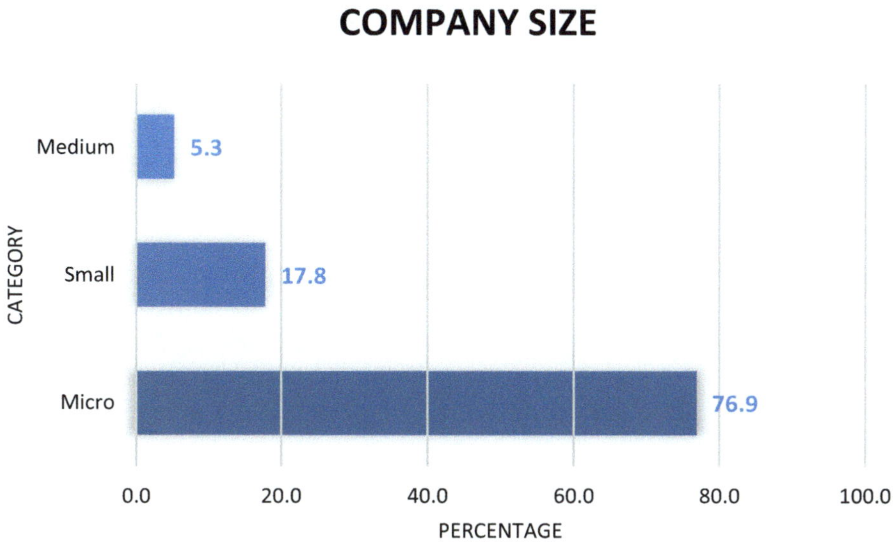

Fig. 8.1 Company size distribution in the sample (N = 398, own data)

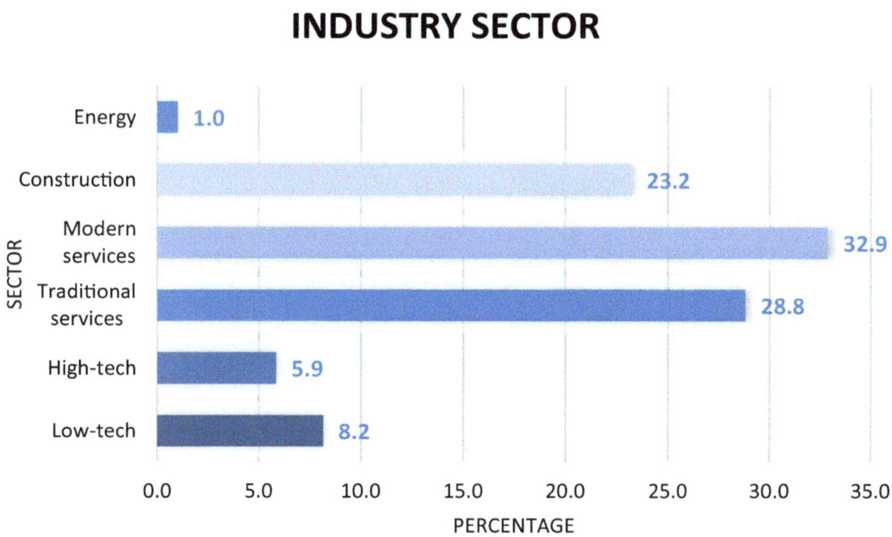

Fig. 8.2 Industry distribution in the sample (N = 392, own data)

As shown in Fig. 8.3, 6% of company managers expect a strong increase in turnover and 42% an increase, which means that in total almost half of the companies in the canton of Schwyz expect to expand their business over the next 2 years. A further 41% do not expect a decline, as they do not anticipate any change from today. Only 11% expect their business to deteriorate. Figure 8.4 shows that most

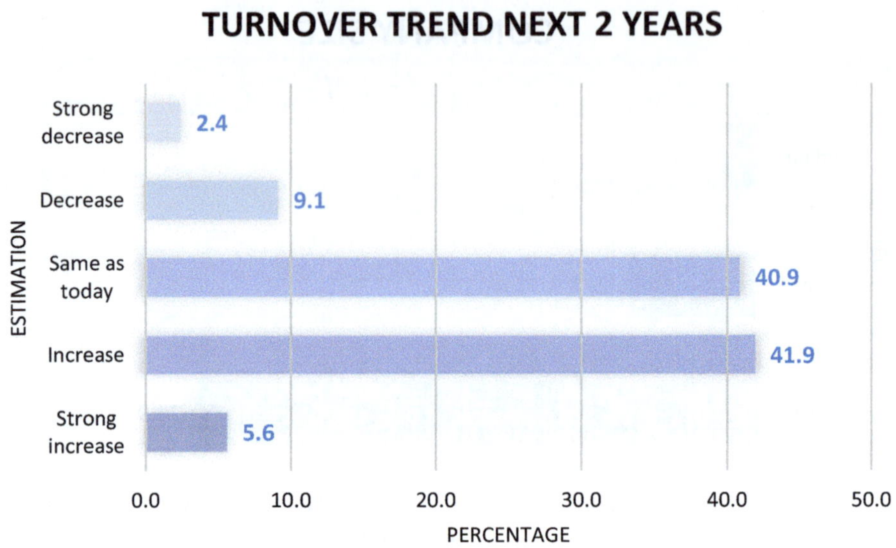

Fig. 8.3 Turnover trend in the sample of companies (N = 372, own data)

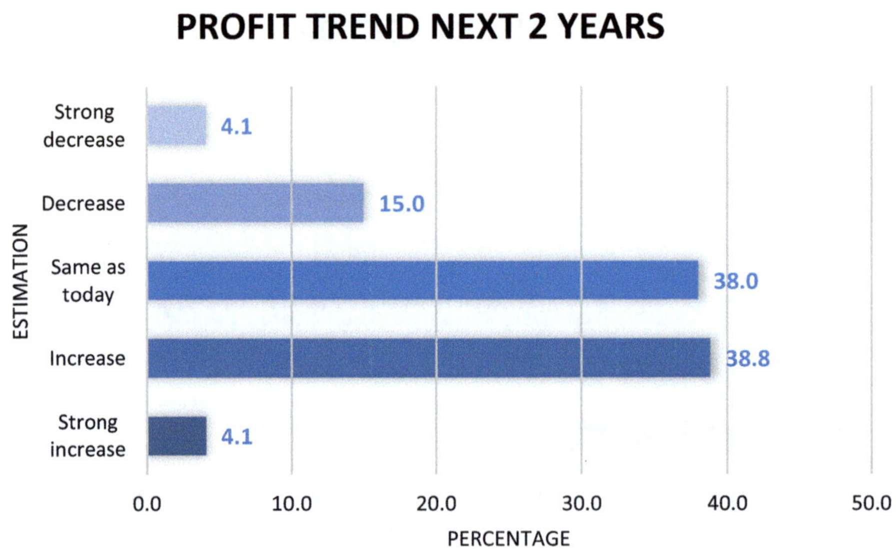

Fig. 8.4 Profit trend in the sample of companies (N = 366, own data)

respondents expect their profits to increase significantly (4%), increase (39%) or remain the same (38%) over the next 2 years. 4% expect a strong decrease in profits and 15% expect a decrease. Overall, the data suggest a generally stable to slightly positive outlook for the next 2 years. However, not all businesses expect to benefit

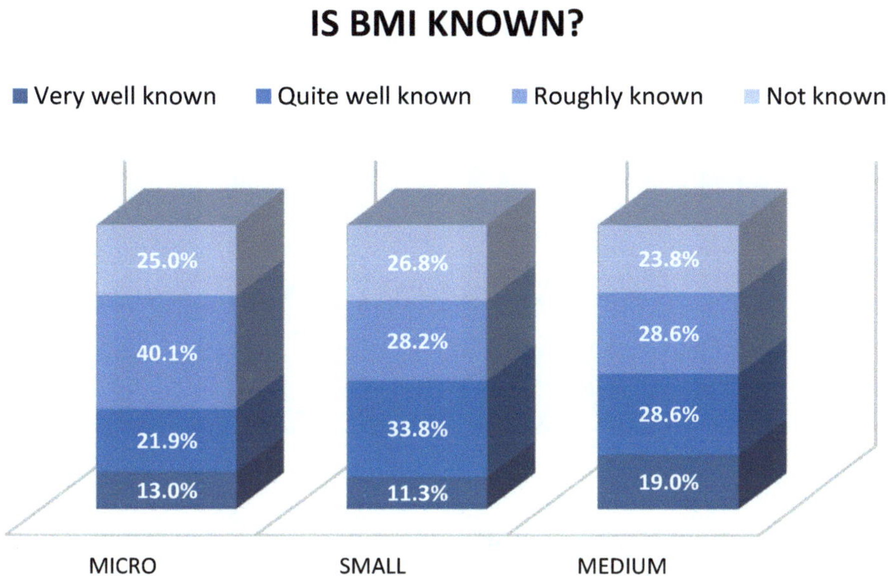

Fig. 8.5 BMI knowledge related to size (N = 384, own data)

from the positive sales situation, as the proportion of those expecting a fall in profits is higher than the proportion expecting a fall in sales.

Figure 8.5 illustrates the BMI knowledge related to size, and Fig. 8.6 shows BMI knowledge related to industry.

Based on the data in Fig. 8.5, in each company size category there are approximately 25% of the company managers who are not familiar with BMI. For the remaining 75%, the proportion of managers that either "very well" or "quite well" know BMI increases with the company size. Figure 8.6 shows the distribution of knowledge of BMI across industries. The proportion of executives who know BMI "very well" or "quite well" is highest in the energy and high-tech sectors. Conversely, knowledge of BMI is lowest in the construction and low-tech sectors. Traditional and modern services have mixed levels of awareness, with traditional services tending to have more limited knowledge.

RQ1: What is the current status of the planned BMI projects?

The data suggest that the likelihood of undergoing a BMI within the next 2 years increases with the size of the enterprise. Small companies are more likely to have no innovation or only a single innovation, such as a product or process innovation. Table 8.5 illustrates the proportion of BMI planned within next 2 years related to size.

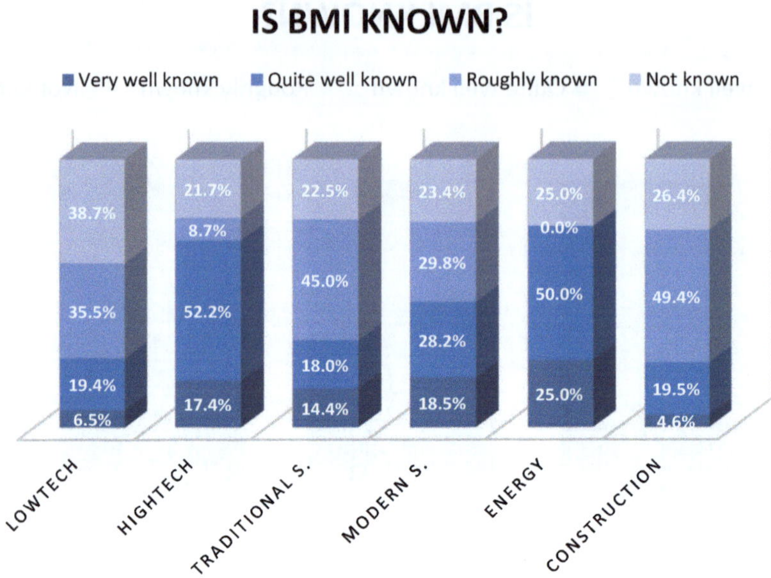

Fig. 8.6 BMI knowledge related to industry (N = 380, own data)

Table 8.5 Proportion of BMI planned within next 2 years related to size [N = 385] (own data)

Will a BMI be done within the next 2 years?	No innovation	Single innovation	Multiple innovations (BMI)
Micro	30.4%	31.7%	*37.9%*
Small	16.9%	26.8%	*56.3%*
Medium	4.8%	4.8%	*90.5%*

As Table 8.6 shows, the high-tech sector has the highest likelihood of undergoing a BMI within the next 2 years, followed by the energy and low-tech sectors. The construction sector has the lowest probability of doing a BMI, but the second highest probability for a single innovation such as a product or process innovation. Table 8.6 shows proportion of BMI planned within next 2 years related to industry.

In this context, it is interesting to see whether companies that have undergone a BMI in the last 10 years are likely to do so again in the next 2 years. Therefore, Spearman's Rho can be used to show the degree of association between two ordinal variables that are not equally spaced, without having to look at the distribution of the data. These variables are BMI performed in the last 10 years and intention to have another BMI in the near future (next 2 years).

Table 8.7 shows the rank correlation (Spearman's Rho) for BMI done in the past related to BMI planned in the future.

Spearman's Rho (Table 8.7) shows a correlation coefficient between these two variables of 0.542, indicating a positive, moderate correlation. This means that as the frequency of having had a BMI in the last 10 years increases, so does the

Table 8.6 Proportion of BMI planned within next 2 years related to industry [N = 381] own data

Will a BMI be done within the next 2 years?	No innovation	Single innovation	Multiple innovations (BMI)
Low-tech	29.0%	22.6%	*48.4%*
High-tech	4.5%	18.2%	*77.3%*
Traditional services	27.9%	28.8%	*43.2%*
Modern services	24.0%	33.6%	*42.4%*
Energy	50.0%	0.0%	*50.0%*
Construction	33.0%	30.7%	*36.4%*

Table 8.7 Rank Correlation (Spearman's Rho) for BMI done in the past related to BMI planned in the future (own data)

Spearman's Rho		Was a BMI done within the last 10 years?	Will a BMI be done within the next 2 years?
Was a BMI done within the last 10 years?	Correlation Coefficient	1	0.540
	Sig. (2-tailed)		0.000
	N	386	385
Will a BMI be done within the next 2 years?	Correlation Coefficient	0.540	1
	Sig. (2-tailed)	0.000	
	N	385	386

probability of having a BMI in the next 2 years and vice versa. The Sig. (2-tailed) value of 0.000 indicates that the correlation is statistically significant. Thus, there is a statistically significant, moderate positive correlation between having done a BMI within the last 10 years and the likelihood of having a BMI being done within the next 2 years.

RQ2: Why was there no investment in more business model innovation in the past?

In the survey, managers were presented with 14 different potential barriers to innovation, each assumed to be independent of the others. They were asked to judge whether or not they could lead to a blockage of innovation-related activities in their companies. Table 8.8 shows the occurrence of a BMI within the last 10 years related to potential BMI barriers.

The following barriers have a significant negative relationship with the outcome, suggesting that organisations with these beliefs are less likely to have conducted a BMI in the last 10 years:

- "We see no need for an adjustment."
- "The risks do not justify the expected benefits."

Table 8.8 Occurrence of a BMI within the last 10 years related to potential BMI barriers (own data)

Logistic Regression: Was a BMI done within last 10 years?	B	S.E.	Wald	df	Sig.	Exp(B)	95% C.I.f Lower	or EXP(B) Upper
We see no need for an adjustment.	***−1.016***	0.248	16.814	1	***0.000***	0.362	0.223	0.589
Our employees are sceptical about new things.	0.162	0.321	0.254	1	0.614	1.176	0.626	2.208
Our teams are too culturally diverse.	***0.784***	0.399	3.869	1	***0.049***	2.191	1.003	4.788
We invest cautiously because of the high costs.	−0.106	0.282	0.141	1	0.708	0.900	0.517	1.564
We do not have the own funds.	0.018	0.288	0.004	1	0.951	1.018	0.579	1.788
The risks do not justify the expected benefits.	***−0.658***	0.294	4.992	1	***0.025***	0.518	0.291	0.922
We lack the expertise in implementation.	***−0.640***	0.316	4.087	1	***0.043***	0.528	0.284	0.981
We do not have the necessary technology.	0.262	0.326	0.648	1	0.421	1.300	0.686	2.461
We are fully occupied with the day-to-day business.	−0.266	0.290	0.839	1	0.360	0.767	0.434	1.354
Our internal processes are complex.	0.326	0.320	1.038	1	0.308	1.385	0.740	2.592
Our internal structure is complex in design.	−0.182	0.430	0.179	1	0.673	0.834	0.359	1.936
Our customer base has a complex structure.	0.111	0.326	0.116	1	0.734	1.117	0.590	2.118
Our customers are sceptical about new things.	0.320	0.309	1.075	1	0.300	1.377	0.752	2.521
The market does not allow any adjustment.	***−1.111***	0.358	9.646	1	***0.002***	0.329	0.163	0.664
Constant	1.221	0.284	18.484	1	0.000	3.390		

- "We lack the expertise in implementation."
- "The market does not allow any adjustment."

The other barriers do not show a significant relationship with the dependent variable of having done a BMI, because their Sig. values are above the conventional threshold of 0.050. The cultural diversity has a significant positive relationship with the outcome, indicating that organisations with diverse teams are more likely to have had a BMI in the past than organisations with a more homogeneous culture. Figure 8.7 illustrates the adjustment barrier related to size, and Fig. 8.8 shows the adjustment barrier related to industry.

Figure 8.7 shows that the perceived need for adjustment of the business model increases with the company size by disagreeing to the statement "We see no need for an adjustment". This suggests that the micro-sized companies are more reluctant

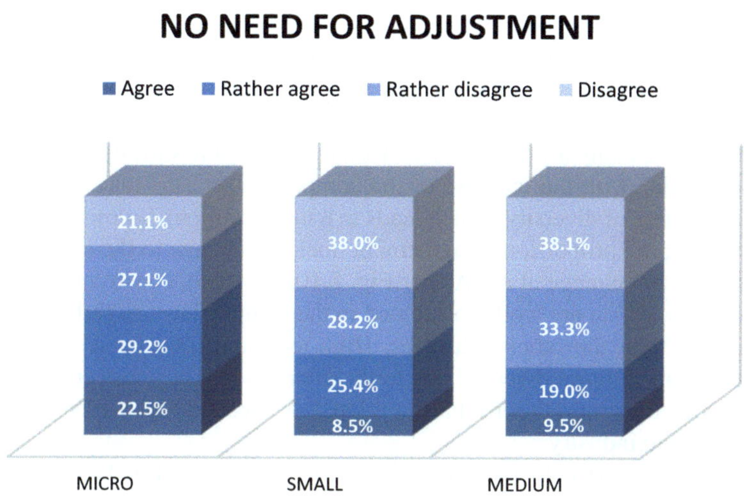

Fig. 8.7 Adjustment barrier related to size (N = 376, own data)

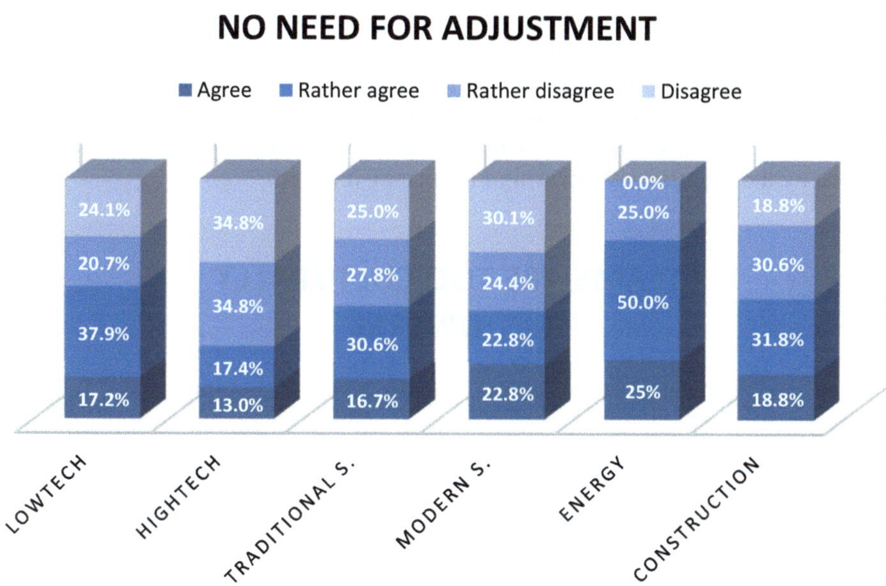

Fig. 8.8 Adjustment barrier related to industry (N = 372, own data)

to innovate their business model, while the medium-sized companies see the value of this activity. As per Fig. 8.8, the high-tech industry has the highest proportion of respondents who either disagree or rather disagree with the statement (69.6% combined), suggesting a more significant perceived need for adjustments in this sector. The modern services industry has the most evenly distributed responses with no clear consensus on the need for adjustments. The low-tech, traditional services, and construction industries have mixed opinions, with a slight majority leaning towards not needing adjustments. Figure 8.9 illustrates the cultural barrier related to size, and Fig. 8.10 shows the cultural barrier related to industry.

Related to the company size (Fig. 8.9), there is a notable shift from "disagree" to "rather disagree" with the increase of the company size. This may be related to a greater awareness of diversity, which leads to a more cautious approach to the question. However, the proportion of agreement does not change to the same extent, as it almost does not change at all. As per Fig. 8.10, low-tech, construction and traditional services, which are well represented in the sample, have slightly higher level of agreement than the other industries. High-tech and modern services may be exposed to more cultural diversity due to the international business environment.

Figure 8.11 illustrates the risk barrier related to size, and Fig. 8.12 shows the risk barrier related to industry.

The belief that the risks do not justify the expected benefits decreases slightly with company size, as the level of disagreement with this statement increases with company size. This implies that organisations that agree with the statement 'The risks do not justify the expected benefits' are less likely to have undergone a BMI in the past, with the micro enterprises being the most reluctant to innovate their business model based on risk-benefit analysis. As shown in Fig. 8.12, this result is similar for the energy, traditional services and construction sectors, as a larger proportion of respondents agree or rather agree that the risks do not justify the expected benefits and therefore the companies have not undergone a BMI in the last 10 years, as implied by the negative statistical significance in Table 8.5.

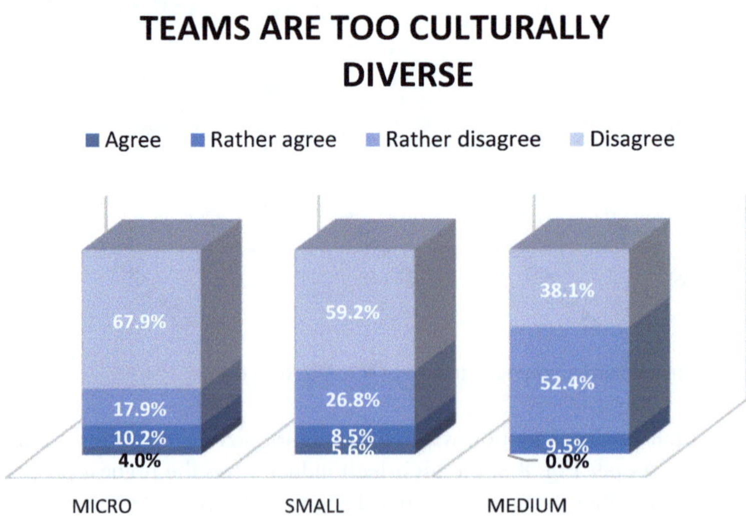

Fig. 8.9 Cultural barrier related to size (N = 366, own data)

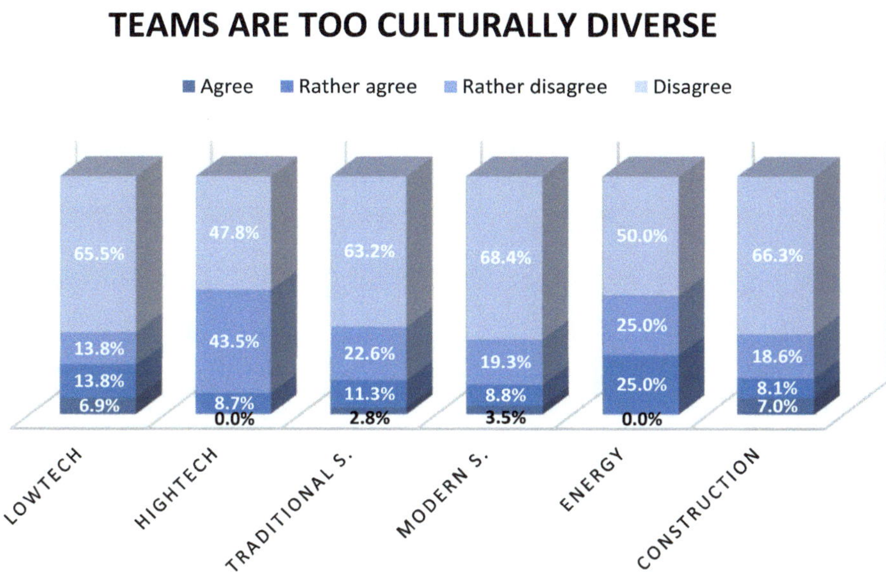

Fig. 8.10 Cultural barrier related to industry (N = 362, own data)

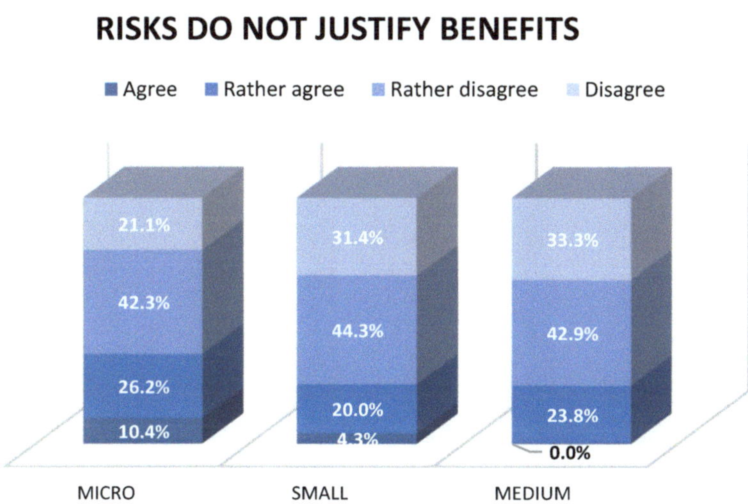

Fig. 8.11 Risk barrier related to size (N = 370, own data)

Figure 8.13 illustrates the expertise barrier to size, and Fig. 8.14 shows the expertise barrier related to industry.

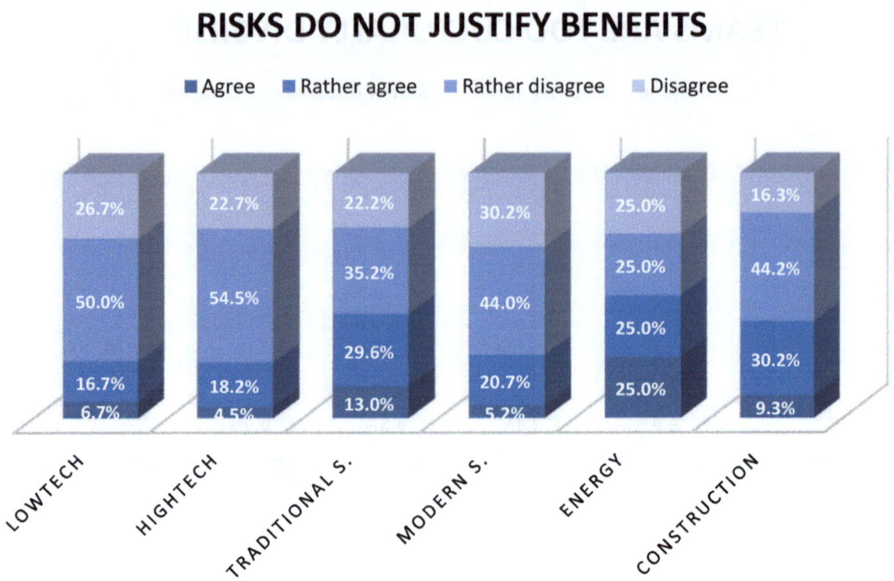

Fig. 8.12 Risk barrier related to industry (N = 366, own data)

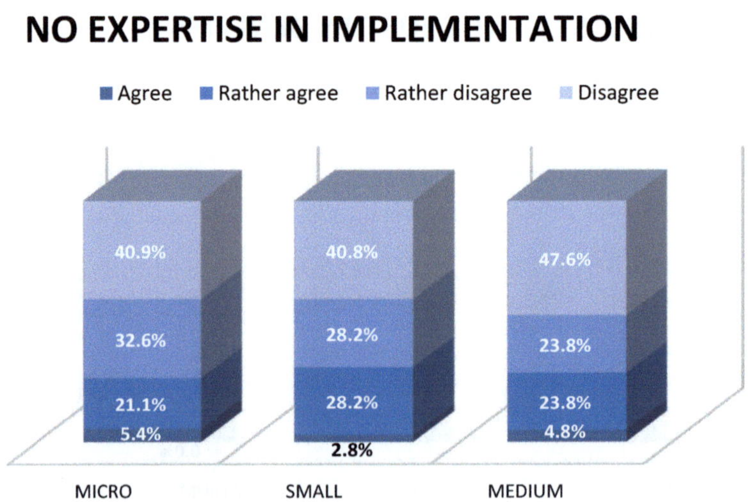

Fig. 8.13 Expertise barrier related to size (N = 371, own data)

Based on the data in Figs. 8.13 and 8.14, it appears that approximately 25–30% of the respondents agree or rather agree to the lack of expertise in implementation across all company sizes and industries, which may overlap to a certain extent with the proportion of respondents not knowing BMI across all industries as per Figs. 8.5 and 8.6.

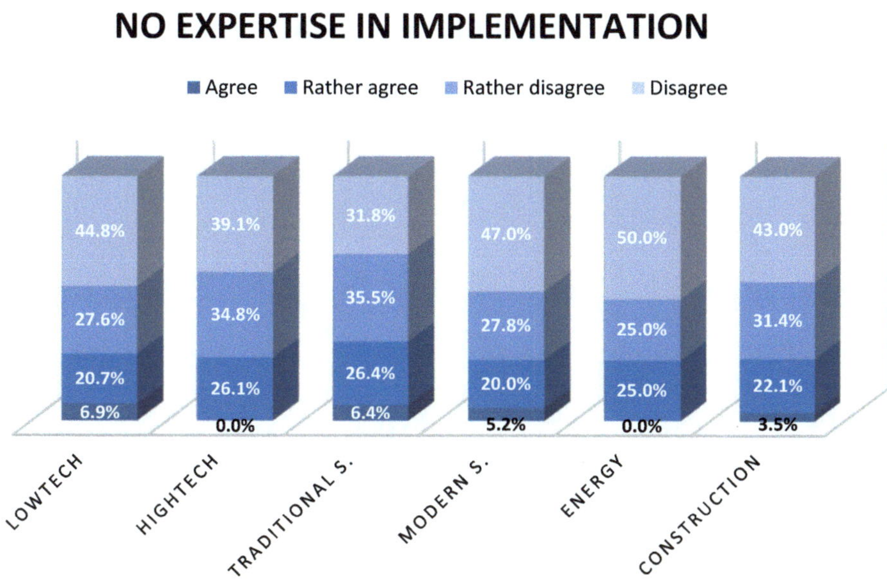

Fig. 8.14 Expertise barrier related to industry (N = 367, own data)

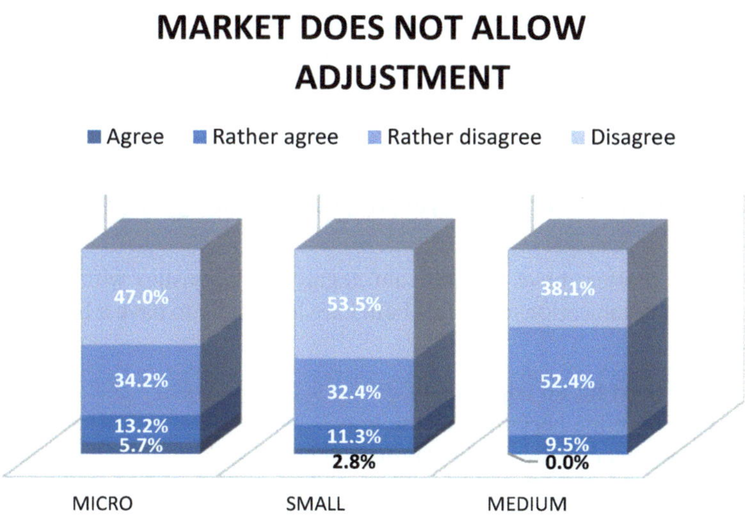

Fig. 8.15 Market barrier related to size (N = 373, own data)

Related to the barrier "The market does not allow an adjustment" the data is as follows in Figs. 8.15 and 8.16:

The data in Figs. 8.15 and 8.16 shows a big proportion of micro-sized companies and traditional service companies that feel limited by the market and are subsequently reluctant to conduct a BMI by expressing agreement (agree and rather

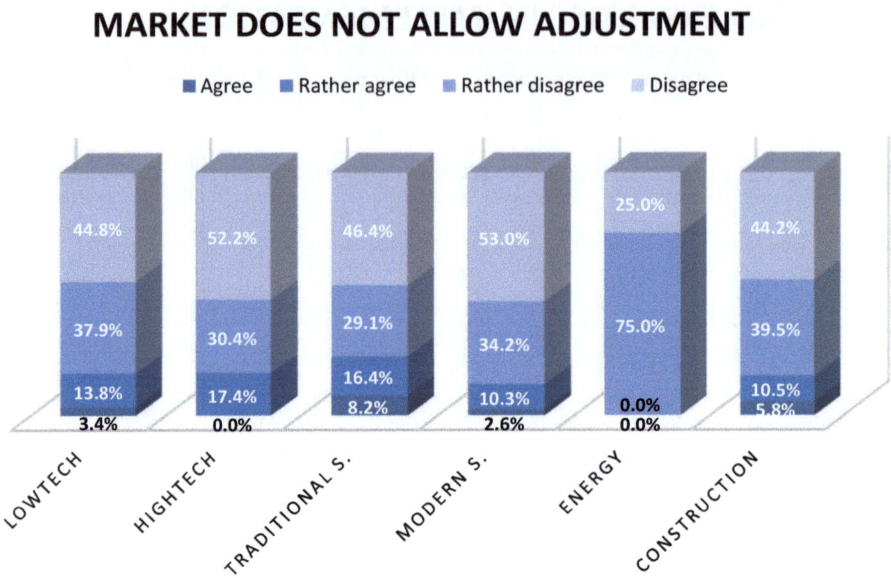

Fig. 8.16 Market barrier related to industry (N = 369, own data)

agree) to the statement "The market does not allow adjustment". Table 8.9 shows significant BMI barriers related to probable occurrence of BMIs within the next 2 years.

RQ3: Are today's barriers different from those of the past?

The following barriers have a significant negative relationship with the outcome, suggesting that organisations with this belief are less likely to have a BMI within the next 2 years:

- "We see no need for an adjustment."
- "We do not have the necessary technology."
- "We are fully occupied with the day-to-day business."
- "The market does not allow any adjustment."

The following barrier has a significant positive relationship with the outcome, suggesting that organisations with this belief are more likely to have a BMI within the next 2 years:

- "Our internal processes are complex."

"We see no need for an adjustment" and "The market does not allow any adjustment" are again identified as significant barriers implying that organisations with these beliefs are less like to have any innovation of their business model within the

Table 8.9 Significant BMI barriers related to probable occurrence of BMIs within the next 2 years (own data)

Logistic Regression: Will a BMI be done within next 2 years?	B	S.E.	Wald	df	Sig.	Exp(B)	95% C.I.f Lower	or EXP(B) Upper
We see no need for an adjustment.	***−1.012***	0.247	16.803	1	***0.000***	0.363	0.224	0.590
Our employees are sceptical about new things.	0.365	0.318	1.319	1	0.251	1.441	0.773	2.687
Our teams are too culturally diverse.	0.153	0.373	0.168	1	0.682	1.165	0.561	2.423
We invest cautiously because of the high costs.	−0.494	0.279	3.124	1	0.077	0.610	0.353	1.055
We do not have the own funds.	0.511	0.289	3.132	1	0.077	1.666	0.947	2.933
The risks do not justify the expected benefits.	−0.079	0.296	0.071	1	0.790	0.924	0.517	1.651
We lack the expertise in implementation.	−0.091	0.322	0.080	1	0.777	0.913	0.486	1.716
We do not have the necessary technology.	***−0.639***	0.323	3.911	1	***0.048***	0.528	0.280	0.994
We are fully occupied with the day-to-day business.	***−0.620***	0.281	4.883	1	***0.027***	0.538	0.310	0.932
Our internal processes are complex.	***0.735***	0.318	5.333	1	***0.021***	2.085	1.118	3.889
Our internal structure is complex in design.	−0.297	0.426	0.485	1	0.486	0.743	0.322	1.713
Our customer base has a complex structure.	−0.263	0.319	0.681	1	0.409	0.769	0.411	1.436
Our customers are sceptical about new things.	0.277	0.302	0.844	1	0.358	1.319	0.730	2.383
The market does not allow any adjustment.	***−0.888***	0.375	5.601	1	***0.018***	0.411	0.197	0.858
Constant	0.924	0.272	11.517	1	0.001	2.519		

next 2 years. In the following the distribution of agreement was investigated for the newly identified significant barriers.

Figure 8.17 illustrates the technology barrier related to size, and Fig. 8.18 shows the technology barrier related to industry.

Figure 8.17 shows that a higher percentage of respondents from larger companies disagreed. This suggests that larger enterprises may have more access to or confidence in their technology than smaller enterprises. As shown in Fig. 8.18, most respondents in the high-tech, modern services and energy sectors disagreed or rather disagreed with the statement, suggesting that they believe they have the necessary technology in their respective sectors. In low-tech, traditional services and construction, opinions are more mixed, with no clear consensus on whether they have the technology to conduct a BMI.

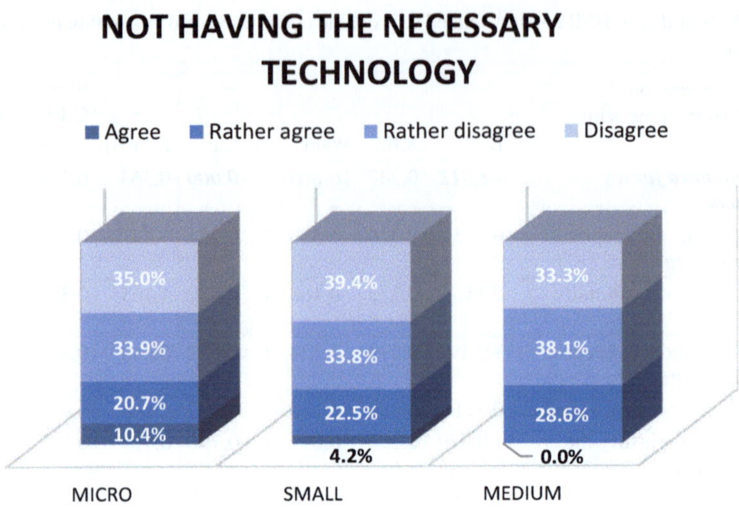

Fig. 8.17 Technology barrier related to size (N = 372, own data)

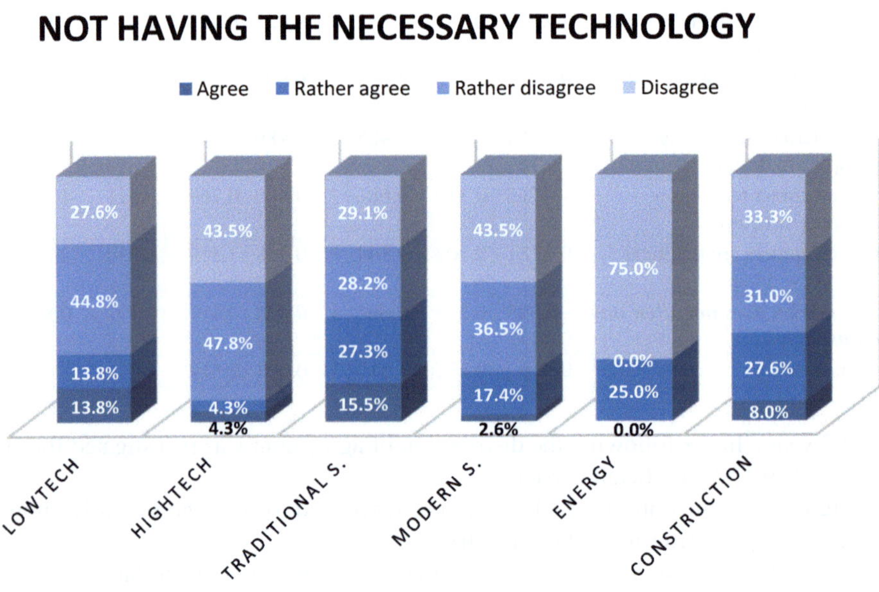

Fig. 8.18 Technology barrier related to industry (N = 368, own data)

Figure 8.19 illustrates the time barrier related to size, and Fig. 8.20 shows the time barrier to industry.

Interestingly, between 70% and almost 80% of respondents across all company sizes (Fig. 8.19) and around 60% to over 80% of respondents across all industries (Fig. 8.20) consider themselves too busy to innovate their business model.

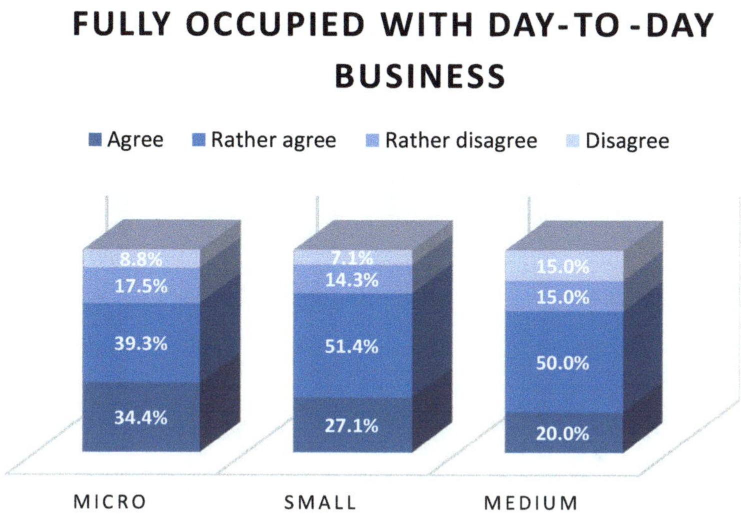

Fig. 8.19 Time barrier related to size (N = 375, own data)

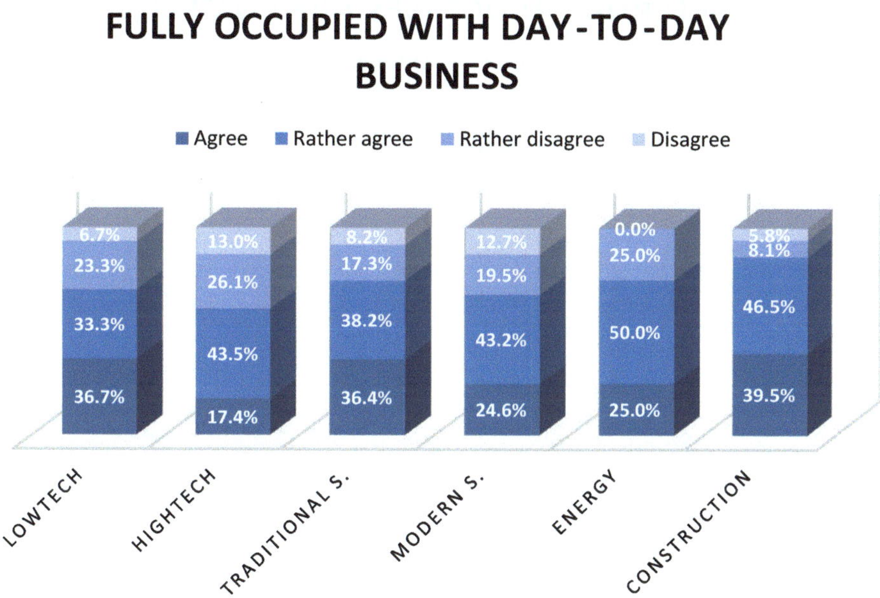

Fig. 8.20 Time barrier related to industry (N = 371, own data)

Figure 8.19 suggests that the level of agreement with being fully occupied with day-to-day business tends to decrease as the size of the company increases. As shown in Fig. 8.20, the high-tech sector has the lowest percentage of respondents agreeing that they are busy, while the construction sector has the highest.

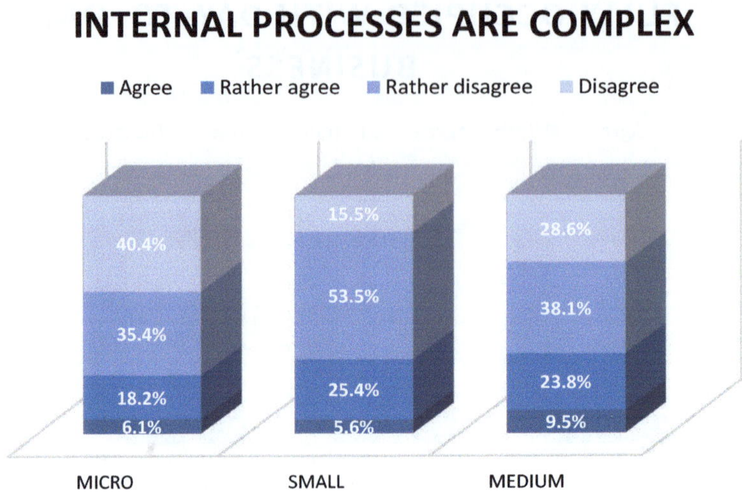

Fig. 8.21 Process barrier related to size (N = 372, own data)

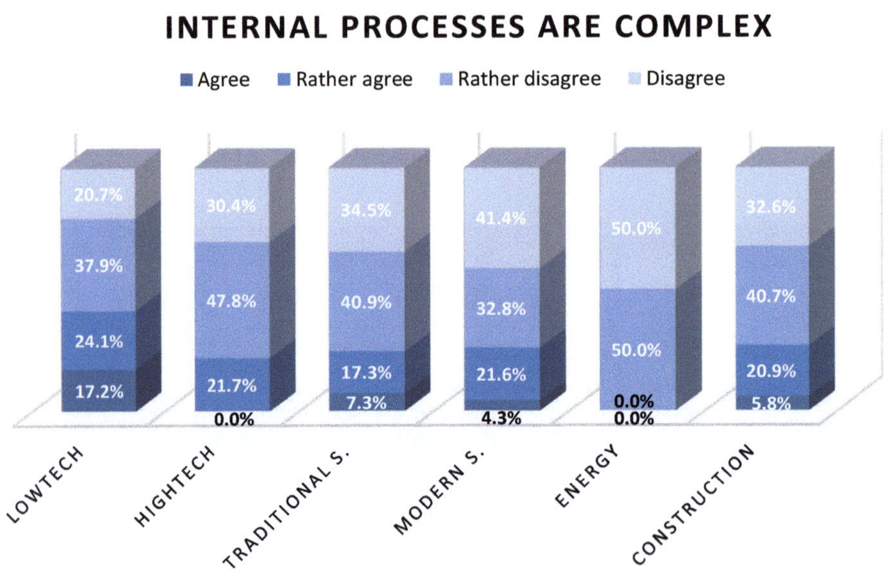

Fig. 8.22 Process barrier related to industry (N = 368, own data)

Figure 8.21 illustrates the process barrier related to size, and Fig. 8.22 shows Process barrier related to industry.

Based on the data in Fig. 8.21, it appears that most respondents across all organisation sizes do not consider their internal processes to be complex. Respondents from micro and medium-sized organisations were more likely to disagree, while those from small organisations were more likely to rather disagree. If the internal processes are seen as being simple, this would explain the positive relationship with a BMI being done within the next 2 years as the fact of having simple internal processes could be seen as an enabler of BMI rather than of a barrier to it. The data in Fig. 8.22 shows that most respondents in the different industries tend to either rather disagree or disagree that internal processes are complex. The low-tech industry had the highest percentage of respondents who agreed or rather agreed that internal processes are complex, implying that they are less likely to undergo a BMI within the next 2 years if simple processes are seen as an enabler of BMI.

RQ4: Does the use of big data, or lack thereof, have an impact on business model development?

Table 8.10 presents the results of the logistic regression analysis of the dependent variable "Is big data being used?". Overall, the findings indicate that the intention to plan a BMI in the next 2 years is significantly associated with current use of big data. Companies planning to innovate their business models in the future are more likely to currently use big data. The trend that companies that had conducted a BMI in the last 10 years are more likely to use big data today is not statistically significant.

Table 8.11 illustrates the carrying out of BMI in the past related to usage of big data.

Table 8.11 demonstrates that the utilisation of big data significantly influenced BMI over the past decade. The positive coefficient (B = 1.029) indicates that as the usage of big data increased, the likelihood of having conducted a BMI also increased. The odds ratio (Exp(B) = 2.797) suggests that the probability of having conducted a BMI is approximately three times higher when big data was used, compared to when it was not utilised.

Table 8.12 shows the planning of BMI in the future related to the usage of big data.

Table 8.12 shows that using big data makes it more likely that a BMI will be done in the next 2 years. The positive coefficient (B = 1.312) indicates that as the usage

Table 8.10 Usage of big data related to doing a BMI (own data)

Logistic Regression: Is Big Data being used?	B	S.E.	Wald	df	Sig.	Exp(B)	95% C.I.for EXP(B)	
							Lower	Upper
Was a BMI done within the last 10 years?	0.533	0.327	2.657	1	0.103	1.704	0.898	3.234
Will a BMI be done within the next 2 years?	***1.046***	0.307	11.602	1	***0.001***	2.848	1.559	5.200
Constant	−3.771	0.557	45.802	1	0.000	0.023		

Table 8.11 Carrying out a BMI in the past related to the usage of big data (own data)

Logistic Regression: Was a BMI done within the last 10 years?	B	S.E.	Wald	df	Sig.	Exp(B)	95% C.I.for EXP(B) Lower	Upper
Is Big Data being used?	***1.029***	0.278	13.691	1	***0.000***	2.797	1.622	4.823
Constant	0.054	0.116	0.216	1	0.642	1.056		

Table 8.12 Planning a BMI in the future related to the usage of big data (own data)

Logistic Regression: Will a BMI be done within the next 2 years?	B	S.E.	Wald	df	Sig.	Exp(B)	95% C.I.for EXP(B) Lower	Upper
Is Big Data being used?	***1.312***	0.268	23.955	1	***0.000***	3.713	2.196	6.278
Constant	−0.488	0.120	16.641	1	0.000	0.614		

of big data increases, the likelihood of conducting a BMI also increases. The odds of conducting a BMI within the next 2 years are nearly 3.7 times higher when big data is used compared to when it is not (Exp(B) = 3.713). In other words, companies should consider using big data if they are planning for substantial innovation of their business model.

Discussion

This study sheds light on the barriers to business model innovation among small and medium-sized enterprises in the Swiss canton of Schwyz. Not surprisingly, the data show that companies that have engaged in BMI in the past, are also more likely to innovate their business models in the future. Also, the high-tech sector having the highest propensity towards BMI does not surprise, as these companies traditionally have to be at the forefront of technological development. In terms of the company size, medium-sized companies are more often engaged in BMI than smaller companies, which is in line with the literature (Spescha & Wörter, 2020).

Over the last 10 years, believing that a BMI is not necessary, being risk-averse, lacking the expertise to carry it out or feeling constrained by market rules have significantly prevented companies from carrying out a BMI, as the relationship with the dependent variable of having carried out a BMI is negative (N). Conversely, companies with a more diverse workforce may have been slightly more likely to have conducted a BMI. The detailed distribution data also show that micro enterprises were more likely to see no need to conduct a BMI, while the high-tech sector was most likely to see a need to conduct a BMI. Larger companies seemed to be more exposed to the risk of conducting a BMI, while the energy, traditional services and construction sectors were the most risk averse. In addition, there was always a proportion of around 25–30% across all sizes and sectors that were unaware of the BMI. Table 8.13 illustrates BMI barriers identified for the past 10 years.

Table 8.13 BMI barriers identified for the past 10 years (own illustration)

Barriers past 10 years
N: We see no need for an adjustment.
P: Our teams are too culturally diverse.
N: The risks do not justify the expected benefits.
N: We lack the expertise in implementation.
N: The market does not allow any adjustment.

Turning to the current plans of the sample enterprises, the size-related result of the past seems to be confirmed by the fact that larger enterprises plan to conduct a BMI more often than smaller ones, while smaller enterprises plan a single dimension innovation, e.g. a new product, or no innovation at all. Across all 6 sectors, the most technical industry (high-tech) seems to be the most likely to plan a BMI. As a rule of thumb, the more technical the sector, the more often a BMI is planned in the next 2 years.

Looking in more detail, the relevant barriers to conducting a BMI in the next 2 years are again the belief that a BMI is not needed or the feeling of being constrained by market rules. Contrary to the previous focus, additional reasons for not conducting a BMI in the next 2 years are technology driven or too busy with routine operations. The perceived complexity of internal processes seems to be more of a trigger for a BMI in the near future due to its positive (P) dependency. The detailed distribution data also shows that larger companies have more access to or trust in technology, as do high-tech companies. Conversely, smaller companies feel more preoccupied with day-to-day business, as do companies in the less technical sectors, such as construction. In addition, complex internal processes seem to trigger a BMI in the future.

Table 8.14 explains the BMI barriers identified for the next 2 years.

Linking this to the use of knowledge/data for the business model, it appears that there is a statistically significant relationship across the companies in the sample between the planning of a BMI in the next 2 years and the use of big data. Even more significant is the reverse situation: the use of big data is a significant factor that has multiplied or multiplied the likelihood of conducting the BMI based on the insights generated. This confirms the findings from the literature (Sorescu, 2017; Ciampi et al., 2021).

In addition to the barriers listed in Table 8.14, the general lack of awareness of BMI was also confirmed as a barrier in the survey results. In line with the literature, the lack of consideration of the customer perspective was not found to be a statistically significant barrier (complex customer base and sceptical customers in Table 8.14), which confirms the observation from the report of the Swiss State Secretariat for Education, Research and Innovation (Spescha & Wörter, 2020) for SMEs, particularly for traditional services and construction, which do not make much use of digital technologies, but are present as a majority in the survey sample. The other negatively connotated barriers (N) were confirmed as statistically significant, in line with the literature. Interestingly, two positively connotated barriers (P) are described as barriers in the literature: cultural diversity (Friedrich von den

Table 8.14 BMI barriers identified for the next 2 years (own illustration)

Barriers next 2 years
N: We see no need for an adjustment.
N: We do not have the necessary technology.
N: We are fully occupied with the day-to-day business.
P: Our internal processes are complex.
N: The market does not allow any adjustment.

Eichen et al., 2015) and complex internal processes (Chesbrough, 2010). In the survey, however, they are positively related to the implementation of a BMI. This means, for example, that the more diverse a team is, the more likely it is to implement a BMI. And the more complex the internal processes, the more likely it is that a BMI will be carried out. Thus, in the context of the survey conducted in the canton of Schwyz, they are seen as a trigger/enabler for BMI rather than an obstacle to BMI. This may be due to the high proportion of modern service companies in the sample.

Interestingly, internal resistance did not have a significant impact, either in the past or in the future. This contrasts with the findings of Chesbrough (2010), who identified resistance to change as the biggest barrier to BMI. Foss & Saebi (2016) also mentioned organisational resistance as a major hurdle, as it threatens the current positions and privileges of employees. We can assume that this may be related to the large size of the SMEs in the sample. In smaller companies it is easier to involve the whole team in innovation processes, whereas in larger companies it is often more top-down. This is why organisational resistance did not have a big impact.

Table 8.15 illustrates the identified significant barriers and anticipated linkage to SME Type, KIS Focus and Data Affinity.

Looking at each barrier and evaluating the initial grouping, the insights generated based on the survey conducted in the canton of Schwyz and in particular on the SME types (Heikkilä et al., 2017) seem to confirm the following:

- "We don't see a need for adaptation" seems to be the main barrier for micro companies, which are generally more reluctant to innovate the business model, and for industries that do not have a high technical exposure.
- "Our teams are too culturally diverse" (enabler) may be an enabler to BMI, particularly for profit-seeking SMEs, as they tend to optimise their costs and back-office processes first. In addition, this may be even more relevant the more knowledge and data-driven an SME is, as a diverse workforce brings in multiple views.
- "The risks do not justify the expected benefits" may be a barrier for growth-oriented SMEs in particular, as they may focus on the customer side of risk assessment first, especially if they are working with data.
- "We lack the expertise to implement" and "We do not have the necessary technology" may initially block start-ups, even more so if they belong to the KIS group and are data affine, due to the impact on the entire business model.
- On the other hand, "We are fully occupied with day-to-day business" may generally block BMI for any company that does not have the resources. Larger

Table 8.15 Identified significant barriers and anticipated linkage to SME Type, KIS Focus and Data Affinity (own illustration)

Linking barriers to SME type, KIS focus and data affinity			
Barrier	SME type	KIS focus	Data affinity
N: We see no need for an adjustment.	No BMI interest	No	No
~~Our employees are sceptical about new things.~~	~~Profitability~~	~~No~~	~~No~~
P: Our teams are too culturally diverse.	Profitability	Yes	Yes
~~We invest cautiously because of the high costs.~~	~~Profitability~~	~~Yes~~	~~Yes~~
~~We do not have the own funds.~~	~~All~~	~~Yes~~	~~No~~
N: The risks do not justify the expected benefits.	Growth	No	Yes
N: We lack the expertise in implementation.	New Business	Yes	Yes
N: We do not have the necessary technology.	New Business	Yes	Yes
N: We are fully occupied with the day-to-day business.	All	No	No
P: Our internal processes are complex.	Profitability	Yes	No
~~Our internal structure is complex in design.~~	~~New Business~~	~~No~~	~~Yes~~
~~Our customer base has a complex structure.~~	~~Growth~~	~~Yes~~	~~No~~
~~Our customers are sceptical about new things.~~	~~Growth~~	~~No~~	~~No~~
N: The market does not allow any adjustment.	Growth	No	No

companies may have a slight advantage here, based on the results above, as they may not have the same level of resource constraints.

- “Our internal processes are complex” (enabler) can support a BMI in profitability seeking SMEs first, especially if they belong to the KIS group, because they focus first on the quality and efficiency of their key activities.
- “The market does not allow any adaptation” may be a barrier to BMI for mainly growth-oriented SMEs, as they focus on value proposition.

For policy makers, it is recommended to implement awareness workshops to demonstrate the importance and benefits of BMI, especially targeting the companies that do not know BMI and do not see need for adaptation. Policy makers can develop workshops to help companies understand and mitigate the risks of innovation. The lack of expertise and technology may be reduced through mentorship programmes, the availability of learning material on tools and techniques and grants for the development of technology, for example together with institutions of higher education. The mentorship programmes can help companies to reorganise their resources to not being fully occupied with the day-to-day business anymore. Furthermore, the exchange of best practices in the context of innovation management and daily business operations can be facilitated through peer learning and networking opportunities. This can also lead to the identification of unrecognised market opportunities that may not have been apparent to managers who feel that the market will not allow them to adapt. In addition, policy makers can provide financial incentives for innovative projects within companies and for innovation hubs or co-working spaces.

Lastly, the regulatory environment should make it as easy as possible for companies to innovate, e.g. through short approval processes.

Conclusion

Besides the knowledge barrier (BMI is not known), 6 further barriers have been identified as relevant barriers to innovation, particularly for BMI, for a sample of companies listed in the commercial register of canton of Schwyz. Two additional barriers turned out to rather be a trigger/enabler for BMI. The barriers taking effect mostly in the past were rather intrinsic, related to diversity, risk averseness and own expertise. The barriers most relevant for the future are cost, technology, and capacity driven. In general, companies not seeing the need or feeling limited by the market rules have not and will not conduct a BMI, even though they may know BMI. This study gives an overview of barriers in Swiss SMEs and can be a starting point for companies that want to innovate to be aware of where the pitfalls lay.

Future Research

Due to the narrow focus of this paper and the wide heterogeneity in the sample, within the service sector particularly, an industry-focused follow-up work, e.g., one for traditional (non-digital) services and a second one for modern (KIS, digital) services may bring further detailed insights on the barriers to BMI. In addition, tailoring future surveys to the SME Type in addition to the industry sector may generate valuable insights on the usage and relevance of BMI in the related SMEs.

Limitations

The response rate of 10% is one of the limitations, which could lead to a potential non-response bias in the results as the participants who responded may not be representative of the entire target population. Additionally, there is a risk of self-selection, where individuals who chose to respond may have other opinions that do not reflect the entire target population, e.g. people that have no interest in innovation might participate less. This could affect the objectivity of the results. Also, the results are cross-sectional and only represent one canton of Switzerland. These factors should be considered when interpreting the results of the study.

References

Acs, Z. J., Audretsch, D. B., & Lehmann, E. E. (2013). The knowledge spillover theory of entrepreneurship. *Small Business Economics, 41*, 757–774. https://doi.org/10.1007/s11187-013-9505-9
Andersen, T. C. K., Aagaard, A., & Magnusson, M. (2022). Exploring business model innovation in SMEs in a digital context: Organizing search behaviours, experimentation and decision-making. *Creativity and Innovation Management, 31*(1), 19–34. https://doi.org/10.1111/caim.12474
Bashir, M., & Verma, R. (2017). Why business model innovation is the new competitive advantage. *IUP Journal of Business Strategy, 14*(1), 7–17.
Bashir, M., Naqshbandi, M. M., & Farooq, R. (2020). Business model innovation: A systematic review and future research directions. *International Journal of Innovation Science, 12*(4), 457–476. https://doi.org/10.1108/IJIS-06-2020-0081
Bereznoi, A. (2015). Business model innovation in corporate competitive strategy. *Problems of Economic Transition, 57*, 14–33. https://doi.org/10.1080/10611991.2014.1042313
Bocken, N. M., & Geradts, T. H. (2020). Barriers and drivers to sustainable business model innovation: Organization design and dynamic capabilities. *Long Range Planning, 53*(4), 101950. https://doi.org/10.1016/j.lrp.2019.101950
Chaudhary, R., Pandey, J. R., & Pandey, P. (2015, October). Business model innovation through big data. In *2015 international conference on green computing and Internet of Things (ICGCIoT)* (pp. 259–263). IEEE. https://doi.org/10.1109/ICGCIoT.2015.7380469
Chesbrough, H. (2010). Business model innovation: Opportunities and barriers. *Long Range Planning, 43*(2), 354–363. https://doi.org/10.1016/j.lrp.2009.07.010
Ciampi, F., Demi, S., Magrini, A., Marzi, G., & Papa, A. (2021). Exploring the impact of big data analytics capabilities on business model innovation: The mediating role of entrepreneurial orientation. *Journal of Business Research, 123*, 1–13. https://doi.org/10.1016/j.jbusres.2020.09.023
Euchner, J., & Ganguly, A. (2014). Business model innovation in practice. *Research-Technology Management, 57*, 33–39. https://doi.org/10.5437/08956308X5706013
European Commission, D.-G. for I. M., Industry, Entrepreneurship and SMEs. (2023). *SME definition*. Retrieved June 20, 2024, from https://single-market-economy.ec.europa.eu/smes/sme-definition_en
Federal Statistical Office. (2024). *Population*. Retrieved June 20, 2024, from https://www.bfs.admin.ch/bfs/en/home/statistics/population/effectif-change.html
Field, A. (2017). *Discovering statistics using IBM SPSS statistics* (5th ed.). Sage.
Foss, N. & Saebi, T. (2016). The bumpy road to business model innovation: Overcoming cognitive and organizational barriers. *The European Business Review*. Retrieved June 20, 2024, from https://www.europeanbusinessreview.com/the-bumpy-road-to-business-model-innovation-overcoming-cognitive-and-organisational-barriers/
Friedrich von den Eichen, S., Freiling, J., & Matzler, K. (2015). Why business model innovations fail. *The Journal of Business Strategy, 36*(6), 29–38. https://doi.org/10.1108/JBS-09-2014-0107
Guldmann, E., & Huulgaard, R. D. (2020). Barriers to circular business model innovation: A multiple-case study. *Journal of Cleaner Production, 243*(118), 160. https://doi.org/10.1016/j.jclepro.2019.118160
Heikkilä, M., Bouwman, H., & Heikkilä, J. (2017). From strategic goals to business model innovation paths: An exploratory study. *Journal of Small Business and Enterprise Development, 25*(1), 107–128. https://doi.org/10.1108/JSBED-03-2017-0097
Huđek, I., & Hojnik, B. B. (2020). Impact of entrepreneurship activity sustainable development. *Problemy Ekorozwoju, 15*(2), 175–183. https://doi.org/10.35784/pe.2020.2.17
Kanton Schwyz. (2023). *Wirtschaftsdaten Kanton*. Retrieved June 20, 2024, from https://www.sz.ch/kanton/wirtschaft/wirtschaftsdaten/wirtschaftsdaten-kanton.html/72-210-94-1966-1949
Lopez, F. J. D., Bastein, T., & Tukker, A. (2019). Business model innovation for resource-efficiency, circularity and cleaner production: What 143 cases tell us. *Ecological Economics, 155*, 20–35. https://doi.org/10.1016/j.ecolecon.2018.03.009

Matzler, K., von den Eichen, S. F., Anschober, M., & Kohler, T. (2018). The crusade of digital disruption. *The Journal of Business Strategy, 39*(6), 13–20. https://doi.org/10.1108/JBS-12-2017-0187

Meyer, R., & Meyer, D. (2020). *Die neuen Selbständigen 2020 – Forschungsbericht*. edition gesowip.

Meyer, R., Meyer, D., & Schmutz, T. (2023). Do SMEs actually know what Business Model Innovation is? Evidence from Switzerland. *European Conference on Innovation and Entrepreneurship, 18*(1), 617–625. https://doi.org/10.34190/ecie.18.1.1737

Morabito, V. (2015). Big data driven business models. In V. Morabito (Ed.), *Big data and analytics: Strategic and organizational impacts*. Springer International Publishing. https://doi.org/10.1007/978-3-319-10665-6_4

Mosig, T., Lehmann, C., & Neyer, A. K. (2021). Data-driven business model innovation: About barriers and new perspectives. *International Journal of Innovation and Technology Management, 18*(02), 2040017. https://doi.org/10.1142/S021987702040017

Philippi, S., & Hinz, A. (2018). Business model innovation in Switzerland: The unused potential. *8th Annual international conference on business & economics (CBE)*. https://doi.org/10.5176/2251-1970_BizStrategy18.116

Philippi, S., Hinz, A., & Kabous, L. (2022). How Swiss start-ups deal with business model innovation. *European Conference on Innovation and Entrepreneurship, 17*(1), 408–415. https://doi.org/10.34190/ecie.17.1.534

Porter, M. E., & Kramer, M. R. (2011). The big idea: Creating shared value. *Harvard Business Review, 89*, 2–17.

Rüb, J., Bahemia, H., & Schleyer, C. (2017). An examination of barriers to business model innovation. In *International Conference on Engineering, Technology and Innovation (ICE/ITMC), IEEE* (pp. 335–350). https://doi.org/10.1109/ICE.2017.8279906

SBFI. (2020). *Forschung und Innovation in der Schweiz 2020*. Staatssekretariat für Bildung, Forschung und Innovation.

Sorescu, A. (2017). Data-driven business model innovation. *Journal of Product Innovation Management, 34*(5), 691–696. https://doi.org/10.1111/jpim.12398

Souto, J. (2015). Business model innovation and business concept innovation as the context of incremental innovation and radical innovation. *Tourism Management, 51*, 142–155. https://doi.org/10.1016/J.TOURMAN.2015.05.017

Spescha, A. & Wörter, M. (2020). *Innovation in der Schweizer Privatwirtschaft – Ergebnisse der Innovationserhebung 2018*. Staatssekretariat für Bildung, Forschung und Innovation SBFI.

Teece, D. (2010). Business models, business strategy and innovation. *Long Range Planning, 43*, 172–194. https://doi.org/10.1016/J.LRP.2009.07.003

Karsten Eichler, Dipl.-Ing., MBA, is an engineer with a degree from the Technical University of Berlin and a postgraduate degree in Business Administration from the University of Applied Sciences Northwestern Switzerland. He started his career as a production worker in the chemical industry before moving on to become a research scientist in the food industry. He has now been working as a consultant and project manager in the pharmaceutical sector for over a decade and a half. Throughout his career, he has held various positions in regulatory, quality, supply chain, development and pipeline operations, often dealing with digitisation, digitalisation and digital transformation. In the context of digitalisation as an innovation driver or enabler, he encounters and effectively manages resistance to change and other barriers to innovation.

Dario Meyer is a researcher at the University of Applied Sciences and Arts Northwestern Switzerland and a PhD candidate in Entrepreneurship at the University of Essex. He holds an MSc in International Management from the University of Applied Sciences and Arts Northwestern Switzerland. He has run his own businesses and worked in several countries in sub-Saharan Africa. His research focuses on entrepreneurship education and training, entrepreneurial ecosystems, and business models.

Rolf Meyer, Prof. Dr., is professor of Entrepreneurship at the University of Applied Sciences and Arts Northwestern Switzerland. He is the head of an MBA-program and researches and consults on business model innovation and entrepreneurship. During his 25 years at the university, he also set up several businesses in various sectors. He holds a doctorate from the University of Basel.

Chapter 9
Integrated Approach to Internationalization in Higher Education

Ivan Köhle and Beat Birkenmeier

Abstract In order to achieve long-term and sustainable success as a company, international markets must be taken into consideration. Universities of Applied Sciences are therefore called upon to take international management topics into account in their curricula. The first part of this chapter documents the results of a study that analyzed the extent to which the business faculties of Swiss Universities of Applied Sciences meet this requirement in their bachelor's degree courses. This part is based on an analysis of publicly accessible module descriptions of the business administration curricula. The results show that three patterns can be recognized: A first group of Universities of Applied Sciences tackle the issue of internationalization by providing degree programs in English with a clear international focus. A second group of universities does not offer a bachelor's degree entirely in English, but offer a significant number of modules related to internationalization, including "compulsory modules", "compulsory elective modules (specialization)", and "other compulsory elective modules. And a third group of universities have no or very few modules with a clear reference to internationalization. In a second part of the paper, the integrated approach of the Internationalization Cube, which was implemented at the University of Applied Sciences and Arts Northwestern Switzerland (FHNW) and its impact is presented.

Keywords Internationalization · Globalization · Education · Universities of Applied Sciences

I. Köhle (✉) · B. Birkenmeier
School of Business, University of Applied Sciences and Arts Northwestern Switzerland (FHNW), Windisch, Switzerland
e-mail: ivan.koehle@fhnw.ch; beat.birkenmeier@fhnw.ch

A. H. Verkuil et al. (eds.), *Core Values and Decision-Making for Sustainable Business*, Sustainable Business Development,
https://doi.org/10.1007/978-3-031-78361-6_9

Introduction

By entering international markets, businesses can diversify their resources and supply chains and reach additional target groups, leading to more resilient operations and subsequently to entrepreneurial success. Besides this financial aspect, internationalization of business fosters global collaboration, allowing companies to share sustainable practices and innovations that address environmental challenges on a global scale. Global business operations encourage the adoption of international sustainability standards, driving companies to commit to more responsible and eco-friendly practices. With international trade, emerging markets benefit and create opportunities for growth, increasing social sustainability with reduces global inequalities.

Internationalization skills are thus prerequisite for the future success of entrepreneurs and managers. Universities around the world are therefore challenged to prepare their students for this new reality of internationalization (Ramaswamy et al., 2021). Furthermore, studies show that internationalization can have a positive impact on students, staff, and faculty (Carrozza & Minucci, 2014; Leask, 2001; Marginson, 2011) through institutional strategies, faculty and student mobility, curriculum, and partnership development (Buckner & Stein, 2020; Stensaker et al., 2019). This is also true for programs on a bachelor's level in the field of business administration. According to Knight (2008), internationalization in higher education can be understood as a process that integrates international, intercultural or global dimensions into the objectives of universities, their areas of responsibility or the services they offer. Internationalization is expressed both in projects and activities as well as in the attitudes and actions of the actors involved (Teichler, 2007; Schröder & Sehl, 2010). For the business faculties of the Universities of Applied Sciences in Switzerland in general and the School of Business of the University of Applied Sciences and Arts Northwestern Switzerland (FHNW) in particular, the topic of internationalization has been a high priority for several years. To date, however, there is a lack of an overview of specific activities of the respective universities. Therefore, a descriptive analysis examines how the business faculties of universities of applied sciences in Switzerland address the topic of internationalization in their bachelor's degree programs and a case study describes the specific approach of the School of Business of FHNW.

The study presented here is part of a series of studies analyzing how the business faculties of Swiss Universities of Applied Sciences and the School of Business at the University of Applied Sciences and Arts Northwestern Switzerland are responding to the challenges associated with social and technological trends. A first study was published by the authors in 2023 on the topic of entrepreneurship (Köhle & Birkenmeier, 2023).

Methods

This study's methodology largely follows the one which was applied by the authors in the study mentioned above (Köhle & Birkenmeier, 2023). The first empirical part of this paper provides a descriptive overview of how Swiss universities of applied sciences with a business faculty deal with the topic of internationalization of their bachelor's degree programs. The study focuses on universities of applied sciences that are members of Swissuniversities. Swissuniversities is the umbrella organization of the Swiss universities, representing them in national and international matters. They promote cooperations and collaborations between various types of higher education institutions, including universities, universities of applied sciences and universities of teacher education. Swissuniversities also represents these universities in international organizations and conferences (Swissuniversities, 2024).

This study analyses the following nine universities of applied sciences, as the profiles of other institutions, such as universities, universities of teacher education and other members of Swissuniversities, are not comparable:

- Bern University of Applied Sciences BFH
- University of Applied Sciences Grisons FHGR
- University of Applied Sciences and Arts Northwestern Switzerland FHNW
- University of Applied Sciences Western Switzerland HES-SO
- Lucerne University of Applied Sciences and Arts HSLU
- Kalaidos Universities of Applied Sciences
- OST—University of Applied Sciences of Eastern
- University of Applied Sciences and Arts of Southern Switzerland SUPSI
- Zurich University of Applied Sciences ZHAW.

The study focuses on bachelor's degrees with 180 ECTS. Master's degrees and Master of Business Administration (MBA) programs are not part of this research. Continuing education programs are also not included in this analysis. This study specifically examines degree programs in business administration. Specifically, it includes bachelor's degree programs in Business Administration, International Management and International Business Management. Other bachelor's degree programs, such as Business Law, Business Information Technology and Business Artificial Intelligence e.g. are excluded due to a lack of comparability.

The study focuses solely on publicly available information on the internet, such as websites and downloadable documents (see references). The analysis was conducted between December 2023 and March 2024 to identify the number of bachelor's degree programs offered in English and/or in collaboration with other countries. Module titles were also searched for keywords such as "internationalization", "globalization", "intercultural" and "cross-cultural".

The modules are classified as "compulsory modules", "compulsory elective modules (specialization)" and "other compulsory elective modules". The "compulsory modules" are normally held during the assessment and are mandatory for all students. For the "compulsory elective modules (specialization)" and "other compulsory elective modules", students have a choice of different modules to attend. These modules are typically only offered as part of the main study program. The differentiation between "compulsory elective modules (specialization)" and "other compulsory elective modules" is based on information provided on university websites. If a clear allocation is not possible, the modules are assigned to the "other compulsory elective modules". The modules generally vary in the number of ECTS credits.

The second part presents the case study of the "Internationalization Cube" which is an integrated approach implemented at the School of Business of FHNW. It places the activities of the School of Business of FHNW in the previously developed classification scheme. The analyses are based on internal documents and personal information.

Results

Internationalization in Bachelor's Programs at Swiss Universities of Applied Sciences

The study demonstrates that all Swiss universities offer modules and language courses in English or other languages, and that studying abroad is generally possible. This indicates that Swiss universities are taking the trend towards internationalization into account. The universities' bachelor's programs can be classified into three groups based on the number of degree programs offered entirely in English, the international cooperation in joint degree programs and the modules listed on their websites (Bern Universtity of Applied Sciences, 2024; Kalaidos Universities of Applied Sciences, 2024; Lucerne Universtity of Applied Sciences and Arts, 2024; OST-University of Applied Sciences of Eastern Switzerland, 2024; University of Applied Sciences Grisons, 2024; University of Applied Sciences and Arts Southern Switzerland, 2024; University of Applied Sciences and Arts Northwestern Switzerland, 2024g; University of Applied Sciences Western Switzerland, 2024; Zurich University of Applied Sciences, 2024):

Group 1: A first group of universities of applied sciences tackle the issue of internationalization by providing degree programs in English with a clear international focus (see Table 9.1).

The School of Business at FHNW offers a triple degree bachelor's program (BSc in International Business Administration) in three languages (French, German, and English) across three countries (France, Germany, and Switzerland). It is important to note that the program is offered in multiple languages and locations. Finally, it is evident that the terms "internationalization", "intercultural", "cross-cultural", and "globalization" are frequently used in conjunction with "management". However, in the category of "compulsory elective modules (specialization)" and "other compulsory elective modules", the number of explicit references to internationalization decreases significantly.

Table 9.1 Universities offering a Bachelor's degree program entirely in English, as well as numerous modules with a clear focus on internationalization

Group 1			
University of Applied Sciences	School of/ Department of	Number of Bachelor's degree programmes (BSc) in Business Administration/ International Management/ International Business Administration entirely in English or in cooperation with other countries.	Modules in the Bachelor's degree programmes (BSc) in Business Administration/International Management/International Business Administration
University of Applied Sciences and Arts Northwestern Switzerland FHNW	Business	1 in English International Management 1 Cooperation with other countries International Business Management Trinational Switzerland, France, Germany	Compulsory modules: • Introduction to International Business • International Taxation • Fundamentals of Cross- Cultural Management • Intercultural Management Styles • Cross-Cultural Leadership • International Economics Compulsory elective modules (specialization): • International Financial Management Other compulsory elective modules: • International Trades
Bern University of Applied Sciences BFH	Business	1 in English International Business Management	Compulsory modules: • International Business Management • Inside the International Firm • International Management and Business Ethics • International Business Law • Environment of the International Firm • Intercultural Competences and Communication Compulsory elective modules (specialization): • Global Management Other compulsory elective modules: • International Taxation • Business and international Policymaking

(continued)

Table 9.1 (continued)

Group 1			
University of Applied Sciences	School of/ Department of	Number of Bachelor's degree programmes (BSc) in Business Administration/ International Management/ International Business Administration entirely in English or in cooperation with other countries.	Modules in the Bachelor's degree programmes (BSc) in Business Administration/International Management/International Business Administration
Zurich University of Applied Sciences ZHAW	Management and Law	1 in English International Management	Compulsory modules: • International Business • International Business and Management • International Environment of Business • Corporate Cultural and Cross-Cultural Management • International Strategic Management • Managing People in an International context • Semester Abroad Compulsory elective modules (specialization): • Not specified Other compulsory elective modules: • Study Trip • International Business Project
Lucerne University of Applied Sciences and Arts HSLU	Business	1 in English International Business Administration	Compulsory modules: • Cultural Competences Compulsory elective modules(specialization): • International Management and Economics Other compulsory elective modules: • International Economics and Strategies for Global Business

Group 2: Another group of universities can be identified that do not offer a bachelor's degree entirely in English but offer a significant number of modules related to internationalization, including "compulsory modules", "compulsory elective modules (specialization)", and "other compulsory elective modules". It is evident that there are numerous modules related to internationalization offered as part of the "other compulsory elective modules". In this category, the average number of listed modules is higher than in Group 1 (Table 9.2).

Table 9.2 Universities that do not offer a Bachelor's degree entirely in English but have a significant number of modules with a clear focus on internationalization

Group 2			
University of Applied Sciences	School of/ Department of	Number of Bachelor's degree programmes (BSc) in Business Administration/ International Management/ International Business Administration entirely in English or in cooperation with other countries.	Modules in the Bachelor's degree programmes (BSc) in Business Administration/ International Management/ International Business Administration
University of Applied Sciences of Eastern SwitzerlandOST	Business	0 International Management 120 ECTS in English	Compulsory modules: • Intercultural Competences • Intercultural Competences and Personal Development • Intercultural Progress Report • International Law Compulsory elective modules (specialization): • International Management • International Business • International Projectmanagement Other compulsory elective modules: • Project Management and International Tools • International Consulting Project 1 • International Consulting Project 2 • International Business Projects
Kalaidos Universities of Applied Sciences	Business	0	Compulsory modules: • International Research Project Compulsory elective modules (specialization): • Not specified Other compulsory elective modules: • International Accounting • International Supply Chain Management • Global Logistics • International Eventmanagement

Table 9.3 Universities that do not offer a Bachelor's degree entirely in English and have few or no modules with a clear focus on internationalization

Group 3			
University of Applied Sciences	School of/ Department of	Number of Bachelor's degree programmes (BSc) in Business Administration/ International Management/ International Business Administration entirely in English or in cooperation with other countries.	Modules in the Bachelor's degree programmes (BSc) in Business Administration/ International Management/ International Business Administration
University of Applied Sciences and Arts of Southern Switzerland SUPSI	Business Economics, Health and Social Care	0	Compulsory modules: • International Business Compulsory elective modules (specialization): • Not specified Other compulsory elective modules: • Not specified
University of Applied Sciences Western Switzerland HES-SO	Business and Services	0	Compulsory modules: • No specified Compulsory elective modules (specialization): • Not specified Other compulsory elective modules: • Global Challenges, SMES & Impact Entrepreneurship
University of Applied Sciences Grisons FHGR	Business and Services	0	Compulsory modules: • Not specified Compulsory elective modules (specialization): • Not specified Other compulsory elective modules: • Not specified

Group 3: The third group of universities have no or very few modules with a clear reference to internationalization. The module titles largely lack any indication of a systematic international orientation (see Table 9.3):

Case Study "Internationalization Cube" at the School of Business of FHNW

The School of Business of FHNW is one of nine schools of the University of Applied Sciences and Arts Northwestern Switzerland (FHNW). It runs three Bachelor of Science degree programs and in 2022, a total of 2580 students studied at Bachelor

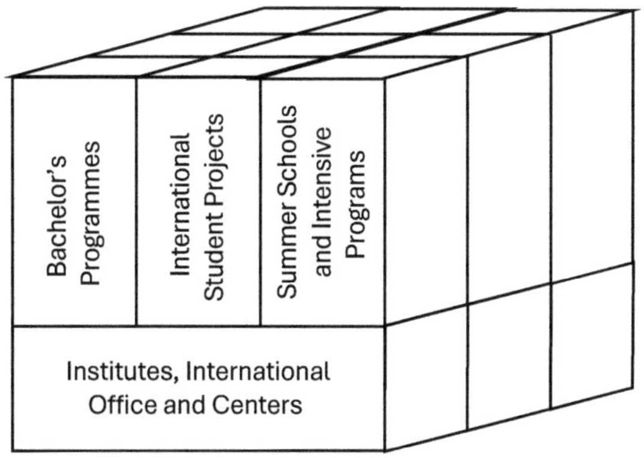

Fig. 9.1 Elements of the "Internationalization Cube" of the School of Business of FHNW

level and in the same year, 549 students graduated with a bachelor's degree (University of Applied Sciences and Arts Northwestern Switzerland, 2023). The school is well networked internationally and fosters around 200 partnerships with other universities around the world (University of Applied Sciences and Arts Northwestern Switzerland, 2024b). In order to make the most of this network and to promote the internationality of its students, the School of Business has established various elements that can be described as the "Internationalization Cube" (see Fig. 9.1).

The "Internationalization Cube" includes structural elements in terms of organizational units on the one hand and different types of activities related to internationalization on the other. All elements can be further subdivided. In the following sections, key variables of the elements of this concept are described in more detail.

Institutes, International Office and Centers The international activities of the School of Business are largely based on the activities of organizational units that are implicitly or explicitly active internationally. These include the six research institutes, whose staff teach the courses, as well as offices and centers that provide dedicated services in connection with internationalization.

Institutes *The research institutes of the School of Business cover the entire field of business administration and are structured in six units: Institute for Information Systems, Institute for Competitiveness and Communication, Institute for Finance, Institute for Nonprofit and Public Management, Institute for Human Resource Management, Institute of Management. Research at these institutes is always geared towards international developments. This is why research projects at FHNW are not just organized within the national context of Switzerland but also with international network partners and are carried out in close collaboration with the industry. Researchers are involved in international applied research and development projects. The benefit of international research funding to the researchers is twofold: some projects receive generous funding and are realized with special expertise. Moreover, researchers are able to build and expand important research networks. In addition to project funding for FHNW researchers, support can also be provided to partners in industry in the context of further European research funding. The benefit* *of these projects and collaborations for education is that the results of these projects can be incorporated into teaching and contacts with international partner universities and companies can be used for joint educational activities.*

International Office The International Office of the School of Business supports students, lecturers and staff in implementing internationalization and is in close contact with international partners. It creates and fosters a range of exchange contracts with partner universities, to enable students, lecturers and staff to gain experience abroad. The intensive contact with government and business bodies in Switzerland, Europe and worldwide makes the School of Business at FHNW a leading, internationally focused School among the Universities of Applied Sciences in Switzerland. Students and staff receive information and advice regarding destinations and study programs. Exchange students from partner universities are informed about entry requirements, academic and student life at FHNW and everyday life in Switzerland. The international office works closely with the students' association and the Erasmus Student Network (ESN) who organize programs and events that bring together incoming and local students and, in this way, foster international collaboration (University of Applied Sciences and Arts Northwestern Switzerland, 2024c).

Latin America Centre The FHNW has retained an involvement with Latin America for over 10 years and is constantly building on its accumulated competencies and networks, particularly in the areas of applied vocational training and culture. Based on its increasingly focused activities in Latin America as well as its long-standing relationships in the realms of science and culture, the FHNW fosters exchange and research cooperation between the Northwestern Switzerland educational area and the Latin American region with a particular emphasis on Colombia (University of Applied Sciences and Arts Northwestern Switzerland, 2024d).

China Centre The China Centre has maintained contacts with Chinese universities and authorities for 25 years. The aim of the China Centre FHNW is to provide Swiss students and companies with profound, up-to-date knowledge of China. The China Centre's activities and services for students, companies and the interested public focus on a well-founded and science-based examination of China's culture, politics and economy. The basis for a critical examination of China and for successful activities with Chinese partners is sound, reflective knowledge of China. The China Centre is known and recognized throughout Switzerland and contributes its China expertise to the China competence map of swissuniversities (University of Applied Sciences and Arts Northwestern Switzerland, 2024e).

Bachelor's Programs The FHNW School of Business currently offers three bachelor's degree programs with a dedicated international focus. An additional one is planned to be launched in autumn semester 2024 (University of Applied Sciences and Arts Northwestern Switzerland, 2024a; University of Applied Sciences and Arts Northwestern Switzerland, 2024e).

Table 9.4 Characteristics of Bachelor's degree programs dedicated to internationalization

Program	Degree	Characteristics regarding internationalization	ECTS points
Bachelor in Business Administration (International Management)	Bachelor of Science FHNW in Business Administration (International Management)—area of specialisation	• Teaching language: EN • Stay abroad (1 semester): mandatory (fulltime track); optional (part-time track)	180
Bachelor in International Business Management (trinational)	International Business Management• Bachelor of Science FHNW• Bachelor of Arts DHBW• Licence Gestion UHA	• Teaching language: EN, FR, DE • 2 semesters in each place: Colmar (FR), Lörrach (DE), Basel (CH) • Stay abroad (1 semester): recommended	210
Bachelor in Business Information Technology	Bachelor of Science FHNW in Business Information Technology	• Teaching language: EN • Stay abroad (1 semester): optional	180

Current programs In addition to the content of the curriculum with subjects such as Introduction to International Business, International Taxation, Fundamentals of Cross-Cultural Management, Intercultural Management Styles, Cross-Cultural Leadership, International Economics, International Financial Management or International Trades, these also have other features of internationalization (Table 9.4).

Future programs In autumn semester 2024 Double Degree (DD) programs at bachelor's level will be implemented for the first time at the FHNW School of Business. Double Degrees broaden students' knowledge in different business subjects and enable them to benefit from a remarkable international learning experience. This program enables students to pursue two academic degrees concurrently from two different universities, allowing them to specialize in a particular field of study or explore a new subject area. Double degree opportunities are open for School of Business FHNW students studying BSc Business Administration (International Management) and BSc Business Information Technology. Students enrolled in these programs can either apply for a Double Degree or complete an exchange semester or take part in any other international opportunity (University of Applied Sciences and Arts Northwestern Switzerland, 2024e).

International Student Projects International student projects all over the world are carried out every year by the School of Business FHNW. Bachelor students in their last year may apply to participate. The projects offer compact short-term programs abroad with an intense framework program and an introduction to the culture and language of the country. All programs are basically open to students from all nine schools of the FHNW and open to incoming students. Each project is organized by a student team and includes a 1-week preparation seminar in Switzerland (lectures, podium discussions and other events) and a 2-week on-site seminar in the

Table 9.5 Projects and destinations belonging to the Intensive programs

Project	Countries visited (2024)
Insight China	China
Focus India	India
ConnectUS	USA, Canada
HOLATAM	Brazil, Argentina
ExploreASEAN	Singapore, Vietnam, Thailand

countries visited (see table). The projects provide a unique opportunity to meet managers, foreign students and leaders from a range of organizations and businesses abroad. During the projects, the students are accompanied by lecturers. A total of around 100 students and 15 lecturers take part in the programs every year. Active participation in a project is rewarded with 5 ECTS credits. Table 9.5 gives an overview over the projects and destinations (University of Applied Sciences and Arts Northwestern Switzerland, 2024g).

Summer schools & Intensive Programs Summer schools and other forms of transnational learning provide an ideal opportunity to study subjects outside the curriculum, gain more in-depth knowledge of certain subjects and build networks (University of Applied Sciences and Arts Northwestern Switzerland, 2024h).

Summer schools The one-to-two-week summer schools may include trips and provide the opportunity to develop language and intercultural competence. In the context of internationalization, the School of Business offers four summer schools. Table 9.6 gives a summary of them:

Every year, around 80 students from the School of Business take part in one of these summer schools with an international focus.

COIL Sustainability Marketing & Ethics. A COIL (Collaborative Online International Learning) is first and foremost a collaboration between two or more universities, between lecturers and between students. The universities are usually located in different countries and the lecturers in a particular subject area, work together and initiate online collaborations between their students. The COIL sustainability Marketing & ethics is part of a co-teaching and co-learning program of FHNW with Sheridan College in Ontario, Canada. Students practice inter-cultural communication whilst digitally interacting with their Canadian peers. The course is embedded in the regular curriculum. Depending on the program students are following, it might belong to the core curriculum or be an optional subject/elective. Thus, students will be awarded credits for it.

Conclusions

In summary, it can be concluded that the School of Business of FHNW offers an above-average intensity of activities in the field of internationalization compared to the other universities of applied sciences in Switzerland. The research shows that

Table 9.6 Summer Schools in the context of internationalization

Program	Target group	Content/Focus	Place	Duration	ECTS points
BSwiss—Swiss International Business Summer School	Swiss and international bachelor's students	Different topics such as sustainability, innovation, migration & competitiveness and culture, politics & competitiveness	Basel	2 weeks	5
Social Entrepreneurship for Economic Development	Undergraduate students from all disciplines	Social entrepreneurship in the context of a developing country in Southeast Asia	Vietnam	2 weeks	5
Summer School EAFIT Colombia – FHNW Switzerland	Swiss and Colombian students from all faculties	Entrepreneurship combining combines profit, non-profit and sustainable projects at the same time	Basel/Medellin	2 weeks	5
Trinational KTUR Summer School Entrepreneurship	Students at all levels in the Upper Rhine region/ international students	International entrepreneuership	Basel/Freiburg/ Strassbourg	1 week	2

around half of all graduates at bachelor's level take advantage of a course specifically geared towards internationalization in the course of their education at the FHNW School of Business. It is evident that the range of courses on offer is constantly being expanded and new internationally oriented courses are being added to the range. With its international orientation, the School of Business FHNW prepares students well for future challenges so that they can contribute to global ecological and social sustainability in addition to economic corporate success.

Limitations

The methodology used allows conclusions to be drawn about the visibility of the topic of internationalization on the websites of the respective universities. However, the visibility is only an indication of the anchoring of the examined topic. The present study is a snapshot without any indication of the development of the topic of internationalization. Neither is it possible to assess the anchoring among students, since the public information does not allow any conclusions to be drawn about the number of students who attend the respective modules. This also means that the present study is not a ranking of the universities in terms of their orientation to the topic of internationalization.

Directions for Future Research

In further research projects, the number of students on international management courses could be surveyed and the impact of teaching on future management tasks analyzed. The study concept could be expanded to other countries' university programs. Furthermore, the connection between internationally oriented management skills and the contribution to sustainability goals could be investigated.

References

Bern University of Applied Sciences. (2024). *Studieren am Departement Wirtschaft* [online]. https://www.bfh.ch/wirtschaft/de/studium.

Buckner, E., & Stein, S. (2020). What counts as internationalization? Deconstructing the internationalization imperative. *Journal of Studies in International Education, 24*(2), 151–166.

Carrozza, C., & Minucci, S. (2014). Keep on Movin'? Research mobility's meanings for Italian early-stage researchers. *Higher Education Policy, 27*, 489–508.

Kalaidos Universities of Applied Sciences. (2024). *Studium* [online]. *https://www.kalaidos-fh.ch/de-CH/Studium*

Knight, J. (2008). Internationalization: Key concepts and elements. In M. Geabel et al. (Eds.), *Internationalization of European Higher Education* (pp. 1–22). Josef Raabe.

Köhle, I., & Birkenmeier, B. (2023). Concept and impact of an integrated approach to entrepreneurship in higher education. In *Proceedings of the 18th European conference on innovation and entrepreneurship*, Part 1, Vol. 18, Nr. 1, pp. 100–109.

Leask, B. (2001). Bridging the gap: Internationalizing university curricula. *Journal of Studies in International Education, 5*(2), 100–115.

Lucerne University of Applied Sciences and Arts. (2024). *Studium - Ein Grundstein für den Berufserfolg* [online], https://www.hslu.ch/de-ch/wirtschaft/studium

Marginson, S. (2011). Higher education and public good. *Higher Education Quarterly, 65*(4), 411–433.

OST - University of Applied Sciences of Eastern Switzerland. (2024). *Wirtschaft, Management und Recht oder Wirtschaftsinformatik studieren* [online], https://www.ost.ch/de/studium

Ramaswamy, M., Marciniuk, D. D., Csonka, V., Colò, L., & Saso, L. (2021). Reimagining internationalization in higher education through the United Nations sustainable development goals for the betterment of society. *Journal of Studies in International Education, 25*(4).

Schröder, T., & Sehl, I. (2010). Internationalisierung von Hochschulen. Ergebnisse eines deutsch-österreichischen Benchmarking-Verfahrens. *HIS: Forum Hochschule, 8*, 1–48.

Stensaker, B., Lee, J. J., Rhoades, G., Ghosh, S., Castiello-Gutiérrez, S., Vance, H., Calikoglu, A., Kramer, V., Liu, S., Marei, M. S., O'Toole, L., Pavlyutkin, I., & Peel, C. (2019). Stratified university strategies: The shaping of institutional legitimacy in a global perspective. *The Journal of Higher Education, 90*(4), 539–562.

Swissuniverities. (2024). [online], https://www.swissuniversities.ch

Teichler, U. (2007). *Die Internationalisierung der Hochschulen. Neue Herausforderungen und Strategien*. Campus.

University of Applied Sciences and Arts of Southern Switzerland. (2024). *Dipartimento economia, aziendale, sanità e sociale* [online], www.supsi.ch/deass

University of Applied Sciences Grisons. (2024). *Studienangebote* [online], https://www.fhgr.ch/studienangebote

University of Applied Sciences and Arts Northwestern Switzerland. (2023). *Statistiken 2022* [online], https://www.fhnw.ch/de/die-fhnw/facts-und-figures/jahresbericht/media/fhnw-statistiken-2022.pdf

University of Applied Sciences and Arts Northwestern Switzerland. (2024a). *Hochschule für Wirtschaft FHNW* [online], www.fhnw.ch/de/die-fhnw/hochschulen/hsw.

University of Applied Sciences and Arts Northwestern Switzerland. (2024b). *Partner Universities* [online], https://www.fhnw.ch/en/about-fhnw/schools/business/international/partneruniversities

University of Applied Sciences and Arts Northwestern Switzerland. (2024c). *International Office* [online], https://www.fhnw.ch/de/die-fhnw/hochschulen/hsw/international-office

University of Applied Sciences and Arts Northwestern Switzerland. (2024d). *The FHNW Latin America Centre* [online], https://www.fhnw.ch/en/international/fhnw-latin-american-centre.

University of Applied Sciences and Arts Northwestern Switzerland. (2024e). *China Centre* [online], https://www.fhnw.ch/de/die-fhnw/hochschulen/hsw/fhnw-china-centre

University of Applied Sciences and Arts Northwestern Switzerland. (2024f). *Double Degrees* [online], https://www.fhnw.ch/en/degree-programmes/business/international-experience/double-degrees

University of Applied Sciences and Arts Northwestern Switzerland. (2024g). *International Student Projects* [online], https://www.fhnw.ch/en/degree-programmes/business/international-experience/international-student-projects.

University of Applied Sciences and Arts Northwestern Switzerland. (2024h). *International opportunities* [online], https://www.fhnw.ch/en/degree-programmes/business/international-experience

University of Applied Sciences Western Switzerland. (2024). *Studium an der HSW-FR* [online], https://www.heg-fr.ch/de/studium/#Studienangebot

Zurich University of Applied Sciences. (2024). *Studium* [online], https://www.zhaw.ch/de/studium

Ivan Koehle, Prof. Dr., has been responsible for the BSc program in Business Administration at the University of Applied Sciences and Arts in Northern Switzerland FHNW, School of Business, since 2010. Ivan Koehle graduated from the University of Zurich and then worked as an assistant at the university's Institute of Accounting. In this position, he worked as a consultant in the field of bank management and completed his doctoral thesis in finance. Before joining the University of Applied Sciences as a study coordinator, Ivan Koehle was co-owner of Kontiki Reisen AG, a travel agency specializing in Nordic destinations. After selling Kontiki Reisen AG to the Kuoni Group, Ivan Koehle managed the integration of a subsidiary into the new group as its managing director.

Beat Birkenmeier, Dr., is a lecturer of Entrepreneurship & Innovation at University of Applied Sciences and Arts Northwestern Switzerland. After studying mechanical engineering as well as industrial and production engineering and working in the transportation industry, he worked as a research assistant and lecturer at the Chair of Technology and Innovation Management at the Institute of Business Administration at the Swiss Federal Institute of Technology (ETH) Zurich. At the same time, he completed his doctorate in the field of management, technology and economics at ETH Zurich. In addition to his research and teaching activities, he is the founder of a Spin-off-Company of ETH Zurich and works as a management consultant with a strong focus on innovation management.

Chapter 10
Second-hand Luxury for Generations Y and Z: Embracing Responsible Consumption or Hunting for Luxury Brand Treasures? An Analysis from France Using the Theory of Planned Behavior

Agnès Walser-Luchesi, Anne-Catherine Furst, and Landisoa Rabeson

Abstract This book chapter focuses on the responsible consumption behavior of young individuals from Generations Y and Z and on the commitment of luxury companies to the sustainability of their products. The objective of this research is to understand the gap between the intention to engage in the responsible consumption of second-hand luxury goods and the actual act of purchasing them. We use the theory of planned behavior to analyze the factors that explain the intentions of younger generations to acquire such goods. The results of our quantitative study indicate that: (1) attitude to second-hand luxury goods, subjective norms and perceived behavioral control have a significant influence on the intention to purchase; (2) the perception of quality associated with second-hand luxury goods has the strongest impact on attitude to them; and (3) the consumption of second-hand luxury goods does not represent a form of pro-environmental behavior within the age group studied. However, purchasing second-hand luxury goods is associated with acquiring high quality, timeless products and the desire to access a luxury brand. This is further influenced by a feeling of treasure hunting.

Keywords Sustainability · Circular economy · Consumer behavior · Second-hand luxury goods · Theory of planned behavior

---

A. Walser-Luchesi (✉) · L. Rabeson
HuManiS Research Center, EM Strasbourg Business School, Strasbourg, France
e-mail: agnes.luchesi@em-strasbourg.eu; landisoa.rabeson@em-strasbourg.eu

A.-C. Furst
Strasbourg, France

A. H. Verkuil et al. (eds.), *Core Values and Decision-Making for Sustainable Business*, Sustainable Business Development,
https://doi.org/10.1007/978-3-031-78361-6_10

Box 10.1

Lana, a marketing student in her final year at a business school, exhibits environmentally conscious behavior typical of her generation. She sorts her waste for recycling and prefers to use her bike or public transportation for her daily commute.

As the end-of-year gala approaches, Lana wants to celebrate her achievement with an elegant handbag. She visits several stores in order to compare different models. She is seduced by a Chanel bag, but it is beyond her budget.

Her friend Lisa, who came across ads on Vinted and Vestiaire Collective, advises Lana to explore these platforms for hidden gems. After several days of searching through a vast selection of second-hand items, Lana finally finds the dream piece—a more affordable Chanel bag.

Lana also learns that this luxury brand is actively seeking environmentally friendly solutions. Motivated by this knowledge, she decides to acquire the desired item. On the day of the gala, Lana posted a photo on Instagram of herself proudly displaying her second-hand luxury handbag. From this first experience, she felt satisfaction and pride, reinforcing her conviction that luxury can also be environmentally responsible …

Nowadays, consumers can acquire goods through a variety of channels other than the traditional physical and digital ones. Second-hand purchasing, upcycling, reconditioned goods, thrift stores, garage sales, flea markets and peer-to-peer selling sites are all distribution formulas that compete with buying new goods. Some of these consumption patterns are deeply rooted in the concept of responsible marketing and offer a variety of sustainable options.

The second-hand product, "distinguishes itself from its new equivalent by the fact that it cannot claim a status of first possession in the life of a buyer" (Roux & Guiot, 2008, p. 65). Once stigmatized as a cheap and functional commodity that met the clothing needs of disadvantaged people, the meaning of second-hand clothing has evolved to become a commodity that is desired and actively sought by all social classes (Milgram, 2004).

Within the textile and ready-to-wear industry, the second-hand market has experienced significant growth over the past two decades, driven by the rise of online shopping and the development of platforms for buying and selling second-hand products, such as Leboncoin, Vinted, and Vestiaire Collective. For some time, consumers have been buying second-hand goods for economic, monetary and limited purchasing power reasons, but today they are also motivated to do so by environmental and ecological causes (Box 10.1). However, in Europe, four million tons of clothing are thrown away annually, and only 20% are recycled.[1] To address these alarming data, selling and purchasing second-hand items has become a solution for reducing the quantity of discarded clothing and limiting environmental pollution. Does this also reflect a change in consumer behavior?

The luxury sector is particularly concerned by this change in consumer behavior, for at least two major reasons. First, given the strong environmental impact of the clothing and textile industry, companies are legitimately committing to integrating environmental sustainability into their production and to encouraging the circular economy.[2] The second-hand market is a means to achieve this goal (see Box 10.2). Indeed, in 2021, the global second-hand market in the world of fashion and luxury was valued at 34 billion euros.[3] In France, this market is worth 7 billion euros, of which 1.16 billion relates specifically to textiles.[4] Second, this market presents a way for the sector to approach Generations Y and Z,[5] who can access luxury through second-hand items. These young people represent a major market segment in the purchasing of fashion items and are considered main players in the second-hand market. Their enthusiasm for second-hand luxury items is motivated by the appeal of the vintage, the associated financial savings, the presence of resale platforms (which they are masters at operating) and their environmental awareness (Tu et al., 2022). Companies are therefore seeking to integrate ethical and environmental considerations into their manufacturing and marketing strategies through corporate social responsibility (CSR) initiatives. Figure 10.1 shows Chinese consumers' "Green habits" by generation in 2023.

Box 10.2

Pursuing sustainability in the luxury industry is a process aligned with the United Nations Sustainable Development Goals and the Fashion Pact signed in October 2020 by many industry players. Luxury companies work on new strategies, with the aim of: promoting new business models that increase product use (second-hand, reselling, renting); implementing alternative solutions that transform used products into new ones (repairing, recycling, upcycling); and creating safe and renewable raw materials (bio-materials).

[1] https://multimedia.ademe.fr/infographies/infographie-mode-qqf/
ADEME, Agence de l'Environnement et de la Maîtrise de l'Energie

[2] Deloitte (2022), Global Powers of Luxury Goods. l

[3] Etude Fevad et KPMG (2022), *la seconde main au premier plan*, septembre 2022
https://fr.statista.com/statistiques/1411749/luxe-seconde-main-occasion/

[4] https://www.journalduluxe.fr/fr/business/seconde-main-rapport-bcg-vestiaire-collective

[5] Gen Y: people born between 1981 and 1996; Gen Z: people born between 1997 and 2010.

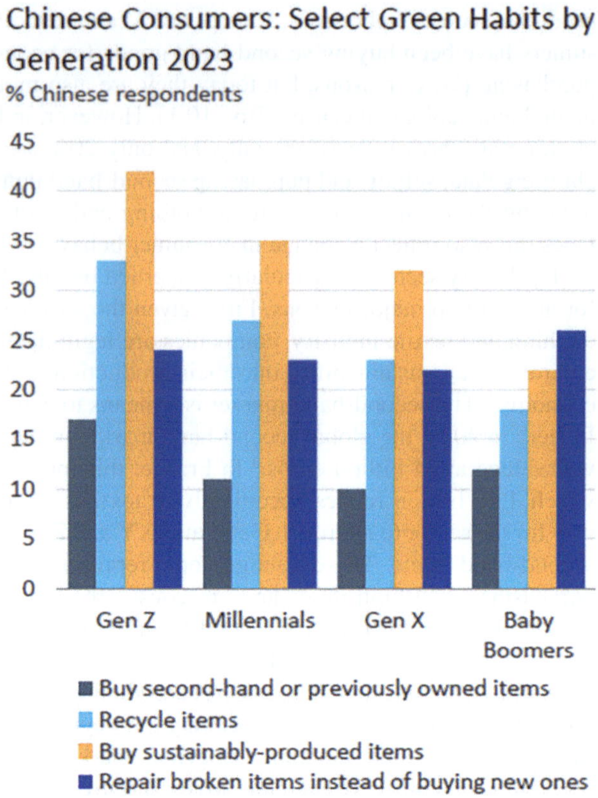

Fig. 10.1 Source: Euromonitor International Voice of the Consumer: Lifestyles Survey 2023

With the growing importance of sustainability issues and conscious consumption, luxury companies have begun to understand the relevance of resale as a strategic solution for limiting the negative impact of their products on the environment. Chinese Gen Zers are more inclined than other generations to acquire second-hand, recycled, repaired and sustainably produced items. This paves the way for reselling, recycling and circular fashion to become more important to Chinese consumers in the near future.

Luxury in the Marketing Literature

To enhance academic knowledge about the purchasing of luxury products, several authors initially defined luxury products as goods of superior quality, expensive and intrinsically rare, that symbolize the status of the purchaser, without being indispensable for daily use (Berry, 1994; Zhang & Cude, 2018). Vigneron and Johnson (2004) added that luxury should be treated as a subjective issue, while being aware of the differences between categories of luxury products. These categories include, for example, high-end, expensive luxury products, such as designer clothes (Shukla, 2010), and accessible luxury products from fashion brands such as Chanel and Hermès. Research has also focused on consumer perceptions of these products based on dimensions of perceived value, including hedonic, altruistic value and a refined taste for personal aesthetics (Ki et al., 2024), alongside utilitarian and functional values.

Research has explored the motivations that drive consumers to buy luxury products. Acquisition of luxury items meets a variety of needs such as the desire for uniqueness. To demonstrate their individuality, an individual with a strong need to be unique will be more inclined to make non-traditional consumption choices, such as buying second-hand clothes (Roux & Guiot, 2008). Other such needs relate to the attainment of a certain status, adoption of a new identity (Kessous et al., 2017), ostentatious consumption (Brodin et al., 2016; Ki et al., 2024; O'Cass & McEwen, 2004), appearance (Freire, 2014; Mick et al., 2004), accomplishment (demonstrating their power and obtaining social recognition, Kessous & Valette-Florence, 2019) and freedom, all encouraged by subjective social norms or the reference group (Mishra et al., 2023; O'Cass & McEwen, 2004). Preeti et al. (2017) highlighted the emergence of social comparison phenomena linked to ostentation and vertical social stratification. Studies have also been conducted in particular contexts or with specific social groups. For example, a luxury product can also be transmitted, thus adding a desire for tradition or heritage (Kessous et al., 2017). For some social groups (subcultures), mimetism is a motivating factor for accessing luxury (Brodin et al., 2016).

Luxury and Responsible Consumption

Few studies have examined how luxury relates to responsible consumption. Those that have done so take two opposing approaches.

In the first approach, sustainability and luxury are considered as two dissonant notions (Fournaise et al., 2023; Kapferer & Michaut-Denizeau, 2019). For instance, the recycled luxury product degrades the rarity and quality of luxury because it is produced from transformed waste or obsolete material (Janssen et al., 2014). In this regard, it can be perceived as deviant behavior of the luxury brand, an unjustifiable act, or can even be considered as a transgression of the luxury brand (Fournaise et al., 2023), positive or negative in the consumer-luxury brand relationship.

In the second approach, the notions of sustainability and luxury are complementary and compatible (Osburg et al., 2021), with common values of superior quality, product longevity, rarity due to the limited availability of materials (Kapferer, 2012) and uniqueness of the product (Hemonnet-Goujot et al., 2022).

Toward a Paradoxical Behavior: The Gap Between Intention to Purchase and the Purchasing Act in Second-Hand Luxury

Despite the growing interest of practitioners and researchers in responsible buying behavior, a paradox has been observed between responsible consumption intentions and the actual purchasing of sustainable products (Carrigan & Attalla, 2001; Moser, 2015; Sempels, 2009), particularly among Generations Y and Z.[6] Few studies have addressed this divergence in the context of second-hand luxury. Intentions translate so weakly into changes in behavior that some authors evoke the myth of the responsible consumer (Carrigan & Attalla, 2001; Eckhardt et al., 2010). The aim of this research is to understand the gap between responsible consumption intentions and the actual act of purchasing second-hand luxury goods, and to identify the antecedents of attitudes to second-hand luxury goods. Drawing on the theory of planned behavior, this research analyzes five factors that influence the attitude of younger generations to purchasing second-hand luxury goods and tests their effects on intention and behavior.

The results of our study, which is based on a quantitative methodology, highlight the factors that determine attitudes to second-hand luxury products. We found that intention to acquire a second-hand luxury product is not a reliable indicator of responsible purchasing behavior and is not necessarily linked to ecological or environmental consciousness. Indeed, purchasing behavior is mainly motivated by the aspiration to own luxury brands for reasons related to personal identity and social success. As luxury brands are no more inclined than younger consumers to perceive second-hand selling and purchasing as a way to engage in the circular economy, this implies that managers of these brands should take steps to make their environmental credentials more visible in order to make their consumers more sensitive to environmental sustainability. This study is also important for resale platforms, particularly in terms of how they communicate to promote the purchasing of second-hand luxury items. It is essential to promote the idea that the environment can be preserved through the sale of second-hand products so that individuals can view this approach as being both economically and ecologically oriented. Major fashion brands often play a key role in conveying this value and, by extension, in promoting sustainability.

Why Choose to Buy Second-hand?

Today, sobriety, cost control and environmental awareness are the main reasons for buying second-hand clothing and accessories (Xu et al., 2014). Therefore, economic motivations from their low prices due to depreciation linked to prior use (Turunen & Leipamaa-Leskinen, 2015) are far from being the only drivers to purchase.

[6] Young people show a strong interest in the environment.

Indeed, psychological (price sensitivity and low level of materialism) and hedonic motivations are also interdependent (Bardhi & Arnould, 2005; Stone et al., 1996) and have evolved.

Researchers have identified additional motivational components, such as the rarity of the product, its geographical origin (Kopytoff, 1986) and a "hedonistic dimension linked, on the one hand, to the designer or luxury products sought and, on the other hand, to a form of serendipity resulting from unexpected encounters with objects" (Roux & Guiot, p. 67). This hedonic motivation[7] is an important driver of second-hand clothing consumption. The fact that second-hand clothing offers consumers the chance to find unique items at affordable (sometimes even negotiable) prices is also associated with an adventurous shopping experience and with the satisfaction of finding something of great value at low cost (Weil, 1999). O'Reilly et al. (1984) highlighted that low price alone was not enough to encourage the purchase of second-hand clothes.

Some researchers have also considered the risks associated with second-hand purchases (Jain & Rathi, 2023), the need to verify authenticity (Ki et al., 2024) and the social pressure that some consumers may feel (Kessous et al., 2017). These factors underline the complexity of second-hand buying behavior and the challenges that consumers face.

Second-hand Purchasing as a Form of Responsible Consumption for Young Generations

Responsible consumption is defined as a set of voluntary actions that consumers take when they are aware of the negative impacts of consumption on the external world (Özçağlar-Toulouse, 2009). In so doing, they consider the social, environmental and national impacts of individual purchases.[1] As consumers are becoming increasingly aware of these impacts, this has led to the emergence of responsible purchasing behavior, which aims to minimize harmful effects and maximize beneficial impacts on society (Mohr et al., 2001).

Generations Y and Z represent a group of consumers who exhibit a strong interest in environmental matters. However, as they do not necessarily have the means to purchase new, environmentally friendly, sustainable products (Schultz et al., 2005), second-hand purchasing makes it possible for them to access these products (see Box 10.3). Growth in the second-hand clothing industry is widespread, and benefits both ordinary and luxury garments. Xu et al. (2014) refer to smart shopping, in which young people are taking a growing interest and where they can use the second-hand channel to find brand-name items at affordable prices and to arbitrate

[7] In the context of our study, the hedonic motivation for purchasing a second-hand luxury product manifests itself through pleasure and avoidance of suffering, here the deprivation of acquiring a luxurious object.

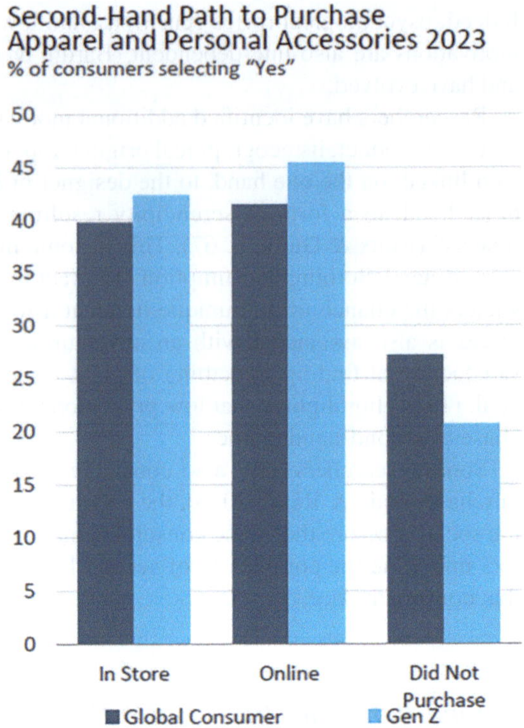

Fig. 10.2 Source: Euromonitor International Voice of the Consumer Lifestyles Survey, 2023

their spending when their budget is restricted. Figure 10.2 shows second-hand path to purchase apparel and personal accessories.

Box 10.3

Consumers are changing how they buy, use and sell luxury goods, and are increasingly interested in the second-hand market. Buying second-hand luxury goods offers consumers a number of key advantages: affordability, ability to collect limited-edition items and collectible editions that preserve their original value. Buying second-hand is a responsible alternative for new generations to acquire items without contributing to the negative impact on the environment. This confirms the value of luxury products that are made to stand the test of time, ensuring that they can remain in circulation over a longer term.

According to Euromonitor Voice of Consumer, Gen Z is the generation most inclined to buy online second-hand items. Thus, luxury goods companies are investing in and partnering with resale platforms as an alternative distribution channel. Resale (online and offline) is the most popular circular economy business model and can include peer-to-peer sales of second-hand items (e.g., on Vestiaire Collective), third-party marketplaces (e.g., RealReal) and own-brand re-commerce.

Circular Economy: A Sustainable Approach to Consumption

The circular economy is an economic model that aims to fight against waste. It encourages the reuse of products by avoiding throwing them away and offering them for resale, donation or second-hand purchase. A circular product is designed to be sustainable, minimizing the consumption of resources and the production of waste. Giving objects a second life allows individuals to possess more items at a lower cost than buying new, while demonstrating responsible behavior. However, this raises the question of whether the objective is to satisfy materialistic goals or to promote environmental responsibility (Bardhi & Arnould, 2005).

Several studies have highlighted the crucial role of consumer awareness of environmental issues, particularly in the luxury sector (e.g., Kapferer & Michaut-Denizeau, 2019). Louis and colleagues (2021) explored the influence of price sensitivity on the gap between intentions and responsible behaviors. However, further investigation is needed to understand how consumers translate this information into personal benefits (François-Lecompte & Gentric, 2016).

The Enthusiasm of Young Consumers for Second-hand Luxury Items

In the specific context of purchasing second-hand luxury products, some studies have examined the motivational factors that most encourage this type of consumption. These factors include the desire for upward social mobility (Kessous & Valette-Florence, 2019), brand attachment, brand desirability and the pleasure derived from the "treasure hunt" experience (Ki et al., 2024). Identifying these factors contributes to a better understanding of responsible buying behavior.[8] However, they do not

[8] Responsible purchasing behavior aims to take account of the social, environmental and national implications of individual purchasing, with a probability of positively influencing these dimensions. Mohr et al. (2001, p. 47) introduced the socially responsible consumer as "a person who bases his acquisition, use and possession of products on the desire to minimize or eliminate harmful effects as well as the desire to maximize the long-term beneficial impact on society".

fully explain the gap between the intention to consume second-hand luxury items and the actual act of purchasing them, nor do they identify the antecedents of attitudes to second-hand luxury goods.

Theoretical Framework and Tested Hypotheses

The theory of planned behavior developed by Ajzen (1991) within the field of social and cognitive psychology has been widely applied in marketing. Its application to explain and predict the emergence of environmentally conscious and ethical consumption behavior in the specific context of second-hand luxury products enables a direct link to be drawn between intention and behavior. By examining attitudes to responsible behaviors (such as buying second-hand, recycling, repairing or reconditioning), subjective norms and perceived control, the theory of planned behavior provides insights into consumers' intentions. Recent research by Stolz (2022) has successfully applied the theory of planned behavior to this context. Stolz's findings highlight that attitude, shopping experiences, perceived behavioral control and subjective norms have a significant influence on purchasing intentions, thus demonstrating the theory's relevance.

To enhance this conceptual framework, scholars have extended the theory of planned behavior model by analyzing factors that impact decisions related to environmentally conscious behaviors (Chen & Tung, 2014; Jaiswal & Kant, 2018). Specifically, considering attitude as a key determinant of purchase intentions for circular products (Modi et al., 2016) and second-hand luxury items (Stolz, 2022), researchers have aimed to identify the antecedents of consumer attitudes. In the theory of planned behavior, attitude is defined as a positive or negative evaluation that occurs when an individual executes a specific behavior (Ajzen, 1991). Taylor and Todd (1995) further state that a positive attitude will lead to a positive behavioral intention, and vice versa. Chen and Tung (2014) posit that consumers' environmental concerns positively influence their attitude toward green hotels, subjective norms, perceived behavioral control and perceived moral obligation, which in turn impact their intention to visit green hotels. In particular, the environmental dimension considered by the authors carries a negative value judgment: environmental concerns reflect consumers' understanding of the detrimental impact on the planet of their purchasing behavior. Our research also focuses on the antecedents of attitude to second-hand luxury products, incorporating the following five factors.

Ecological Motivation: In response to growing awareness of the limited availability of the earth's resources, consumers are increasingly inclined to rethink how

---

Responsible consumption can be defined as "all voluntary acts located in the sphere of consumption carried out following awareness of the consequences deemed negative of consumption on the outside world" (Özçağlar-Toulouse, 2009).

Furthermore, a circular product is designed in a sustainable way, minimizing the consumption of resources and the production of waste.

they consume (Désaunay, 2021). Jan (2022) points out that this awareness of the ecological impact of products directly influences their purchasing intention. Numerous studies, including that by Mandaric et al. (2022), have shown that consumers' attitudes to sustainability when purchasing clothing are strongly linked to their overall concern for social and environmental well-being, their perception of sustainable fashion and their own ethical behavior. As a result, consumers of luxury, aware of the importance of sustainability, feel that it is their responsibility to engage in environmentally and socially responsible consumption practices. In a context of growing ecological awareness, the consumption of second-hand goods appears to be an effective way to encourage recycling, limit the overproduction of objects and, consequently, prevent waste (Guiot & Roux, 2010). As environmental concern has been considered as an antecedent of the components of the extended theory of planned behavior model (Chen & Tung, 2014), to better understand the intentions of Generations Y and Z to buy second-hand luxury, ecological sensitivity will also be considered as an antecedent.

H1a. Ecological motivation positively influences attitudes to purchasing second-hand luxury items.

Knowledge of Brands' Environmental Commitment: Several companies have chosen to communicate their commitment to sustainability, often highlighting comprehensive sustainability reports (Kapferer & Michaut-Denizeau, 2014). Corporate social responsibility is a source of added value for luxury brands, as it has a positive impact on their image in contemporary society (Chen et al., 2022). Informed individuals are more likely to adopt a favorable attitude to second-hand luxury products. Moreover, according to social ideal theory, individuals consider the existence of an ideal society (Rawls, 1999). They care about the well-being of society and actively seek ways to contribute to it (Robeyns, 2008), including acting in an environmentally sustainable way. Thus, individuals with in-depth knowledge of the environmental issues being addressed by companies in the luxury sector are more likely to display a positive attitude to second-hand luxury goods, motivated by their aspiration to actively contribute to the creation of an ideal society.

H1b. Knowledge of brand commitment to sustainable development positively influences attitudes to purchasing second-hand luxury items.

Environmental knowledge and ecological motivation are two positive and ecologically favorable components that lead to environmentally conscious purchasing decisions in favor of environmental protection. This sensitivity can be found in the awareness that consumers have of brands' commitments to the environment and their ecological motivations.

Economic Dimension: Influence of Lower Price on Attitude to Second-hand Luxury Items

The lower price of second-hand products is often cited as a primary reason for choosing them (Cervellon et al., 2012; Turunen & Leipamaa-Leskinen, 2015). Research has considered various aspects related to price, including the desire for affordability, the pursuit of fair pricing, the pleasure of hunting (bargain hunting) and the gratifying aspect of obtaining a product at a low price (Guiot & Roux,

2010). It has also considered the potential resale value, with second-hand luxury products seen as investments for the future (Turunen & Pöyry, 2019).

H1c. The lower price of second-hand luxury products positively influences attitudes to purchasing second-hand luxury items.

Identity Dimension (Dubois, 1994; Holman, 1980; Jain, 2019; Tu et al., 2022; Vigneron & Johnson, 2004): One of the incentives for ostentatious consumption is the consumer's need to display a higher social status (Jaikumar et al., 2018). The purchase of certain products can be a way for consumers to express their self-image (Zeithaml, 1988). This self-projection influences individuals' perceptions and their choice of brands (Holman, 1980).

H1d. Self-image positively influences attitudes to purchasing second-hand luxury items.

High Quality: Quality, exceptional craftsmanship, attractive aesthetics, pleasant texture (Tu et al., 2022), resistance to damage thereby extending their lifetime, and absence of harm to the environment are all characteristics of luxury products. According to Turunen and Leipamaa-Leskinen (2015), the exceptional quality of luxury products makes them suitable for customers who acquire them second-hand or third-hand. Therefore, quality is expected to have a positive influence on attitudes.

H1e. The high quality of second-hand luxury products positively influences attitudes to purchasing luxury items.

Another Variable, the Hedonic Dimension: Weil (1999) highlighted the adventure and excitement associated with searching for high-value products at low cost. Xu et al. (2014) pointed out that individuals search for good deals in thrift stores and second-hand shops. However, O'Reilly et al. (1984) highlighted that low price alone was not enough to encourage the purchase of second-hand clothing. Hedonic motivation (Amaldoss & Jain, 2005; Goldman, 1999; Wattanasuwan, 2005) is essential to explain second-hand consumption (Roux & Guiot, 2008). For instance, other authors have shown that the popular leisure activity of "treasure hunting" (Tu et al., 2022) motivates the purchasing of second-hand luxury items (Kessous & Valette-Florence, 2019) for a variety of reasons. These include nostalgia, recreation, the pursuit of a good deal and seeking excitement (Turunen & Leipamaa-Leskinen, 2015). This treasure hunting gives consumers rewarding experiences and a sense of accomplishment, along with satisfaction from making cost savings, all of which contribute to positive perceptions and attitudes to second-hand luxury products. Consumers driven by this hedonic dimension will modify their attitude and purchasing intentions accordingly. Treating treasure hunting as a moderating variable in the model helps us to better understand the generation under study.

H2. Treasure hunting moderates the relationship between the attitude to purchasing and the intention to buy second-hand luxury items.

Figure 10.3 shows the research model of the authors.

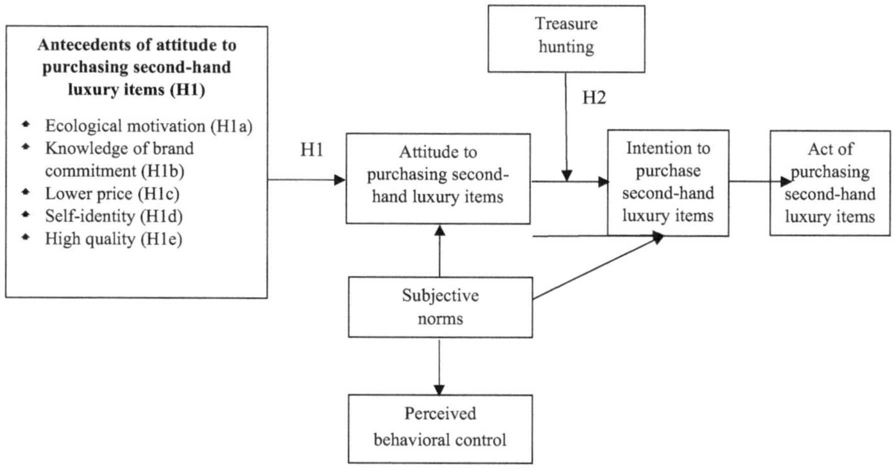

Fig. 10.3 Research Model (own illustration)

Main Results and Discussion

Our empirical investigation followed a rigorous operating procedure (Box 10.4).

Box 10.4: Survey Methodology

*Data collection was carried out in April 2023. The online survey specifically targeted French consumers from Generations Y and Z who engaged in circular economy practices and were familiar with luxury brands. Before answering the questions, participants were subjected to a selection process which aimed to verify their suitability for the desired profile. They were asked two filter questions to identify whether they belonged to Generations Y or Z, and to check whether, after reading the definition, they practiced the circular economy. We opted for a structured, self-administered questionnaire because it generally has higher response rates and limits researcher interference (*Oppenheim, 1992*). The use of online surveys has the advantage of reducing the desirability bias often associated with face-to-face interviews, particularly when it comes to sensitive topics such as environmentally friendly behavior (*Holbrook and Krosnick, 2010*). We surveyed students from the University of Strasbourg, work colleagues and relatives. This approach allowed us to obtain a diversity of opinions and profiles to enrich our study. A total of 210 responses were obtained, of which 165 were considered valid.*

The questionnaire took an average of seven minutes to complete. It was designed to assess the participants' status as consumers and buyers of second-hand luxury items and covered a number of themes such as the practice of second-hand circuits, second-hand luxury, beliefs (about the consequences of their behavior, the expectations of social referents concerning their behavior, obstacles and facilitators) and the respondent's information sheet. The questions were taken from pre-existing scales in the literature, which we adapted to the specific context of our research.

Measures

We used a quantitative research design for this study, developing a survey that included the constructs described in the proposed model. Participants were asked to indicate the degree to which they agreed or disagreed with elements of the survey using a Likert scale with five possible responses (ranging from 1, "strongly disagree," to 5, "strongly agree").

Data Analysis

To analyze the proposed theoretical model and test the hypotheses, we conducted structural equation modeling using SmartPLS v.4.0.0 (Ringle et al., 2015). This approach allowed us to test the entire model and has the advantage of being well suited to exploratory research and accommodating moderate sample sizes. This method also provides greater statistical power compared to maximum likelihood estimators (Benitez et al., 2020).

Model Fit Evaluation

We calculated the standardized root mean square residual (SRMR) as an approximate measure of model fit for PLS-SEM (Henseler et al., 2015). The SRMR showed a value of 0.068, which is below the threshold of 0.08, indicating a favorable fit for PLS-SEM (Henseler et al., 2015).

Convergent Validity Assessment

Next, we tested the reliability and validity of the construct. The saturation of all elements in the final measurement model exceeded 0.707 (Carmines & Zeller, 1979). This confirms the validity and reliability of the measurement scales. Both Cronbach's alpha (Nunnally, 1978) and composite reliability (Chin, 1998) exceeded the minimum threshold of 0.70 (Fornell & Larcker, 1981). In addition, all latent variables successfully achieved convergent validity, as indicated by their average variance extracted (AVE) exceeding the threshold of 0.50 (Fornell & Larcker, 1981).

Furthermore, we assessed discriminant validity using the approach recommended by Fornell and Larcker (1981), specifically examining the heterotrait-monotrait (HTMT) correlation ratios. We observed that the HTMT values were below the threshold of 0.85, indicating satisfactory discriminant validity (Henseler et al., 2015).

Structural Model Evaluation

We assessed collinearity issues using variance inflation factors (VIF). Our analysis indicated that the VIF values in our sample ranged from 1.000 to 4.150, all well below the established threshold of 5 (Hair et al., 2017). Therefore, our model does not exhibit any evidence of multicollinearity.

Results

Our study examined whether second-hand purchasing truly reflects responsible consumption, specifically in the context of luxury second-hand products (Table 10.1). Our findings are as follows:

(1) **Among the antecedents of consumers' attitudes to second-hand purchasing, ecological motivation has no significant effect on their attitude to purchasing second-hand luxury items.** Purchasing second-hand luxury goods does not necessarily imply alignment with ecological and environmental causes. Therefore, for younger generations, the intention to acquire a second-hand luxury item is not a predictor of responsible buying behavior. Instead, it appears to be primarily motivated by the desire to access high quality luxury fashion brands.

Table 10.1 Statistical results (own data)

Effects	Standardized path coefficient (t-Statistics) [95% Confidence Interval]			
Direct effects				
H1a: Ecological motivation → ATT	0.055	(0.810)	[0.081;	0.191]
H2a: Knowledge → ATT	0.160*	(2.495)	[0.038;	0.290]
H3a: Economic motivation → ATT	0.021	(0.286)	[0.114;	0.176]
H4a: High quality → ATT	0.334***	(4.348)	[0.178;	0.480]
H5a: Self ID → ATT	0.162*	(2.361)	[0.021;	0.293]
ATT → INTENT purchase	0.235***	(3.574)	[0.094;	0.350]
Norm → INTENT purchase	0.178**	(7.643)	[0.355;	0.600]
Norm → ATT	0.292***	(4.500)	[0.167;	0.423]
Norm → CONTROL	0.397***	(5.748)	[0.263;	0.536]
Control → INTENT purchase	0.578***	(10.314)	[0.467;	0.689]
INTENT purchase → ACT purchase	0.643***	(12.871)	[0.541;	0.733]
Moderating effect				
H2: Treasure hunting x ATT → INTENT PUR	0.077*	(2.219)	[0.006;	0.146]

Note: ***$p < .001$, **$p < .010$, *$p < .050$; t-values (one-tailed test) are presented in parentheses; 95% bootstrap confidence intervals are presented in [brackets]

(2) **Economic motivation does not influence consumers' attitudes to purchasing second-hand luxury items.** In price theory, authors have demonstrated that price sensitivity is a multidimensional construct with various expressions. Among these, the concept of lower prices does not hold true. This logic aligns with the luxury context, where consumers prioritize quality even when considering second-hand items.

Table 10.1 shows the statistical results of the authors.

(3) Our research shows that perceived **quality** takes precedence over other antecedents in shaping attitudes to purchasing second-hand luxury goods (as also emphasized by Dubois et al., 2001; Kapferer & Bastien, 2009; Vigneron & Johnson, 2004). This finding suggests that the more consumers are aware of the high quality of second-hand items, the more advantages they perceive from buying second-hand luxury products.

(4) The study identifies two other factors that positively predict attitudes: **self-identity**, and **knowledge of the environmental actions** taken by luxury brands. This result is in line with François-Lecompte and Valette-Florence (2006), who emphasized the importance of information in socially responsible consumption. We also show that lack of information is a major barrier to second-hand purchasing in the luxury context.

(5) Our findings demonstrate the moderating effect of **treasure hunting** on the relationship between attitude and intention to purchase second-hand luxury items. This indicates that such shopping experiences can positively influence the intention to buy second-hand luxury products. Through the act of price comparison, consumers discover good deals in a way that is similar to treasure hunting.

Thus, the hedonic dimension outperforms the economic dimension related to low prices in the context of second-hand luxury. This moderation sheds light on individuals' interest in the shopping experience, as Ki et al. (2024) recently emphasized, without explicitly revealing the specifics of moderation. The intention to purchase a second-hand luxury item is enhanced when consumers associate the pleasure of bargain hunting with buying second-hand luxury goods. Consequently, purchasing second-hand luxury items transcends mere transactions and becomes an exciting quest in which each acquisition represents a new form of treasure hunting, where hidden stories and values associated with these products are uncovered.

(6) Our results highlight the positive impact of **social norms** on attitude and behavioral control. Thus, we demonstrate that attitudes to luxury purchases are strongly influenced by societal and peer pressure to buy second-hand items.

(7) We validate the positive influence that the intention to purchase second-hand luxury items has on actual behaviors, going against the paradox found by previous research (Moser, 2015). We also reaffirm the use of the theory of planned behavior based on attitudes to purchasing second-hand luxury items, subjective norms and perceived control. Consequently, our results also show that the intention to purchase a second-hand luxury product predicts actual buying behavior. Overall, our findings confirm the positive influence that the intention to purchase second-hand luxury items has on real-world behaviors.

Conclusion

In the context of second-hand luxury purchasing behavior, the perception of quality takes precedence over environmental consciousness (Kessous & Valette-Florence, 2019). This aligns with the findings of Ki et al. (2024), who emphasize the excellent quality of second-hand luxury items. However, ecological motivation does not significantly influence consumers' attitudes to purchasing second-hand luxury goods.

Self-identity also plays a role in shaping attitudes to second-hand luxury purchasing. This suggests that the more sensitive consumers are to their self-identity, the more advantages they perceive from buying second-hand luxury products. The question remains: Why does the enthusiasm for second-hand luxury purchases not necessarily translate into support for ecological and environmental causes?

When considering the growing interest of consumers of luxury in sustainable issues related to the purchase of long-lasting and timeless products, we need to distinguish between two populations: **(1) traditional consumers of luxury** (who do not necessarily turn to the second-hand market because they have sufficient money to buy new products); **and (2) emerging buyers in the second-hand market**, who represent a new and unexploited target audience for luxury brands, particularly among first-time buyers.

Consequently, the relationship between the intention to purchase a second-hand luxury item and environmentally responsible behavior appears less significant in the context of identity-oriented consumption and within this specific segment of the population. For younger generations, the intention to acquire a second-hand luxury item is not a predictor of responsible buying behavior; instead, it seems to be mainly motivated by the desire to access luxury fashion brands.

In summary, although the second-hand market is a perfect illustration of product reuse and sustainability, there is insufficient visibility of luxury brands participating in this circular economy. If they communicated more about how they contribute to the circular economy, they could improve their image and combat their reputation as polluters. In other words, neither luxury brands nor young consumers are subscribing to the circular economy concept, and this suggests there is potential for luxury brands to communicate more about their environmental actions in terms of CSR, in order to positively reinforce their image.

Managerial Implications

In the emerging context of the second-hand luxury market, it is crucial to analyze the managerial implications for the various stakeholders, including traditional luxury brands, online platforms and specialty stores. To encourage wider adoption and to promote a positive image, particularly among Generations Y and Z, luxury brands need to focus on two major strategic axes: CSR and communication strategy.

It is essential that luxury brands re-examine their perception of responsible consumption and adjust their marketing strategies accordingly (see Box 10.5). This could involve the creation of eco-labels and environmental labeling, as some authors recommend (François-Lecompte & Gentric, 2016), or responsible, ethical labels (François-Lecompte & Valette-Florence, 2006).

Luxury brands can communicate their commitment to sustainable practices by creating labels, as in the following example. Collaborating together, the second-hand resale platform Vestiaire Collective and the luxury brand Alexander McQueen have created a label that aims to ensure the quality and circularity of items. This collaboration entails placing a Brand Approved label on items that Alexander McQueen sources from its customers following evaluation by experts from both brands.[9] Only items approved by both brands are marked with Near Field Communication (NFC) tags containing authentication information accessible to the buyers, which they can access through little chips that allow them to see digital information on their smartphone. This partnership between a resale site and a luxury brand also gives customers assurance about the history and authenticity of the items they purchase. In addition, purchasing items from sustainable companies is a responsible alternative way to acquire items without having a negative impact on the environment.

Figure 10.4 shows the reasons to invest in sustainability for fashion in 2023.

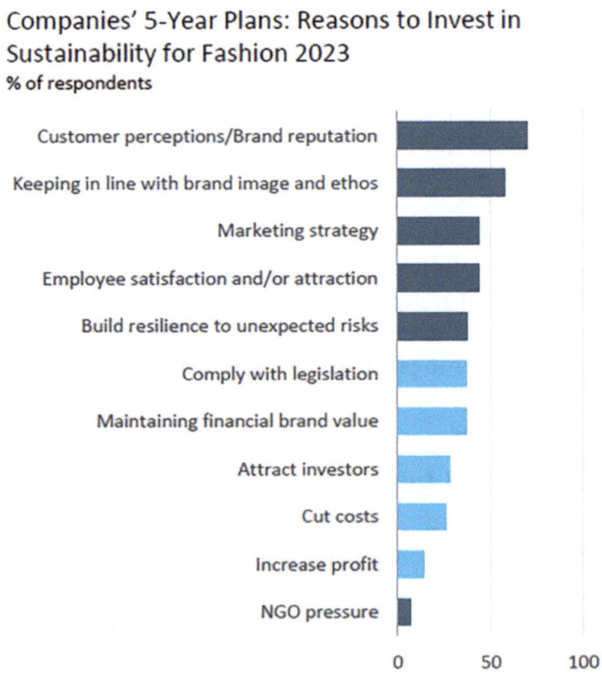

Fig. 10.4 Source: Euromonitor International Voice of the Industry: Sustainability Survey, 2023

[9] https://faq.vestiairecollective.com/hc/en-us/articles/360017709477-Brand-Approved-Alexander-McQueen

Box 10.5

According to Euromonitor International Voice of the Industry, more than 50% of fashion professionals cite "Customer perceptions/Brand reputation" and "Keeping in line with brand image and ethos" as the main reasons for their company's investment in sustainability. The same survey shows that nearly 40% of respondents stated that one of the reasons behind their sustainable investments was "Marketing strategy".

Company management can no longer ignore convergence of the social and environmental dimensions with its economic dimension. Our study has shown the importance of communicating with consumers about companies' environmental actions in order to influence attitudes to pre-owned luxury items. Among other benefits, this enables companies to improve their brand reputation. Investing in the resale market for luxury goods thus gives them a competitive advantage and new market potential.

There is a significant lack of communication by companies about their environmental initiatives, limiting consumers' environmental awareness and the ability to integrate this dimension into the formation of consumer attitudes. Kapferer and Michaut-Denizeau (2014) suggest that luxury companies perceive environmental sensitivity as a key element, but not as a fundamental value. Major fashion brands therefore have an important role to play in transmitting this value and, by extension, in promoting sustainability.

Luxury brands and resale platforms should encourage their customers to help the planet by reducing waste and keeping goods in circulation for as long as possible. For Mother's Day, the second-hand resale platform the RealReal has launched the initiative *Future heirlooms: Mother's Day gift "She'll love them now, You'll love them later"*. In this way, the brand is communicating that sustainability can reaffirm the idea that a luxury item is an investment, a product that never loses its quality, tradition and exclusivity.

It is important to promote the idea of preserving the environment through the sale of second-hand products, so that people perceive this approach to be both economic and ecological. Luxury brands should also take account of the recreational motivations of second-hand (online) store visitors. Every visit should represent a treasure hunting experience, in which the customer can make an incredible discovery that they can describe as a treasure hunting success. Luxury companies should offer a shopping experience that utilizes a state-of-the-art, easy-to-navigate and aesthetically pleasing secure web portal which allows access to a diverse collection of designers offering ultra-luxurious limited editions that are rare and fashionable items, while also being affordable. As luxury items have a history and precious quality, luxury brands should create narratives and stories about items that have traveled through time, highlighting how quality persists over time.

Online platforms and specialty stores also play a key role in promoting the second-hand luxury market. They must facilitate access to second-hand products by offering user-friendly, secure platforms for transactions, while increasing consumer awareness of the importance of responsible purchasing. Creating search filters for sustainable and second-hand products could help consumers to make more conscious purchasing choices.

The RealReal is a large online marketplace for authenticated, second-hand luxury goods, with more than 20 million members. It provides a safe and reliable platform for consumers to buy and sell their luxury items. Items undergo a rigorous authentication process supervised by hundreds of experts and brand authenticators. The RealReal aims to support the circular economy by giving new life to pieces from luxury brands such as Stella McCartney through its Circular Economy Program.[10]

Indeed, Stella McCartney encourages its customers to participate in the circular economy through its partnership with the RealReal. This brand is one of the first luxury brands to join forces with resale platforms as an alternative distribution channel. For example, its alliance includes an incentive for shoppers, who receive $100 to shop at Stella McCartney stores if they consign any Stella McCartney item to RealReal. Similarly, shoppers who buy Stella McCartney items through the second-hand retailer will be rewarded with $100 of store credit to use at Stella McCartney retail stores or online.

Luxury brands should therefore embrace the fact that a strong secondary market supports the primary market. Luxury brand companies should understand how pre-owned items can help to extend the lifetime of their products and increase the relevance of brands among new, mainly young consumers.

By adopting a CSR approach and an appropriate communications strategy, stakeholders can guide consumers toward more responsible and sustainable choices, while consolidating their market position. The circular economy offers opportunities for all stakeholders who can integrate sustainability and ethics into the heart of their global strategy.

In conclusion, the managerial implications underline the importance of a proactive approach to CSR and communication along with the need to create synergies between the various players involved to shape a more sustainable future for the pre-owned luxury market.

Limits and Future Research

There are some limits to this research, particularly with regard to the sample size and the lack of Generation Y participants (only 35% of the sample represented

[10] https://investor.therealreal.com/news-releases/news-release-details/realreal-and-stella-mccartney-strengthen-partnership/

Generation Y, while 65% were from Generation Z). Future research could therefore focus more on analyzing the significant differences between Generations Y and Z, providing a more comprehensive understanding of the distinct behaviors, preferences and motivations of each generation.

To better understand the intention to buy and the act of buying second-hand luxury goods, future research could investigate the role of other variables: the emotional dimension (can emotions such as guilt or pride motivate consumers to buy second-hand luxury goods?); the cultural dimension such as Hofstede's long-term orientation vs short-term orientation (are consumers with a long-term orientation more inclined to buy second-hand luxury goods than those with a short-term orientation?); and, as subjective norms play an important role, future research could also examine the influence of celebrities on the purchasing decisions of young consumers for second-hand luxury goods.

Future research could also explore another aspect of the circular economy—the rental of luxury items—to understand how luxury can be enhanced through sharing, while preserving its inherent notion of exclusivity.

Finally, while we considered the effect of lower price on purchasing attitudes, we could also examine the act of comparing prices and whether the level of intensity of price differentials between new and pre-owned items can change behavior in purchasing second-hand luxury goods.

References

Ajzen, I. (1991). The theory of planned behavior. *Organizational Behavior and Human Decision Processes, 50*, 179–211.

Amaldoss, W., & Jain, S. (2005). Pricing of conspicuous goods: A competitive analysis of social effects. *Journal of Marketing Research, 42*(1), 30–42.

Bardhi, F., & Arnould, E. J. (2005). Thrift shopping: Combining utilitarian thrift and hedonic treat benefits. *Journal of Consumer Behavior, 4*(4), 223–233.

Benitez, J., Henseler, J., Castillo, A., & Schuberth, F. (2020). How to perform and report an impactful analysis using partial least squares: Guidelines for confirmatory and explanatory IS research. *Information and Management, 57*(2), 103168.

Berry, C. J. (1994). *The idea of luxury: A conceptual and historical investigation* (Vol. 30). Cambridge University Press.

Brodin, O., Coulibaly, D., & Ladwein, R. (2016). Le luxe ostensif sous-culturel comme processus mimétique créatif: le cas des sapeurs parisiens. *Revue et Applications en Marketing, 31*(1), 46–66.

Carmines, E. G., & Zeller, R. A. (1979). *Reliability and validity assessment*. Sage.

Carrigan, M., & Attalla, A. (2001). The myth of the ethical consumer – do ethics matter in purchase behaviour? *Journal of Consumer Marketing, 18*(7).

Cervellon, M. C., Carey, L., & Harms, T. (2012). Something old, something used: Determinants of women's purchase of vintage fashion vs second-hand fashion. *International Journal of Retail and Distribution Management, 40*(12), 956–974.

Chen, M.-F., & Tung, P.-J. (2014). Developing an extended Theory of Planned Behavior model to predict consumers' intention to visit green hotels. *International Journal of Hospitality Management, 36*, 221–230.

Chen, L., Halepoto, H., Liu, C., Yan, X., & Qiu, L. (2022). Research on influencing mechanism of fashion brand image value creation based on consumer value co-creation and experiential value perception theory. *Sustainability, 14*, 7524.
Chin, W. W. (1998). The partial least squares approach to structural equation modeling. *Modern Methods for Business Research, 295*(2), 295–336.
Désaunay, C. (2021). Vers la déconsommation? *Constructif, 59*(2), 30–34.
Dubois, B. (1994). Typologie de consommateurs ou de situations de consommation. *Décisions Marketing, 2*, 85–87.
Dubois, B., Laurent, G. & Czellar, S. (2001). Consumer rapport to luxury: Analysing complex and ambivalent attitudes, *Les Cahiers de Recherche*, Groupe HEC: 736.
Eckhardt, G. M., Belk, R., & Devinney, T. M. (2010). Why don't consumers consume ethically? *Journal of Consumer Behaviour, 9*(6), 426–436.
Fornell, C., & Larcker, D. F. (1981). Evaluating structural equation models with unobservable variables and measurement error. *Journal of Marketing Research, 18*(1), 39–50.
Fournaise, T., Kessous, A., & Valette-Florence, P. (2023). When luxury brands use recycled materials: Crossed views between consumers and professionals on a transgressive effect. *Décisions Marketing, 2*, 57–84.
François-Lecompte, A., & Gentric, M. (2016). L'étiquetage environnemental des produits et services: enjeux et défis. *Décisions Marketing, 1*, 99–113.
François-Lecompte, A., & Valette-Florence, P. (2006). Getting to know the socially responsible consumer better. *Décisions Marketing, 41*, 67–79.
Freire, N. A. (2014). When luxury advertising adds the identitary values of luxury: A semiotic analysis. *Journal of Business Research, 67*, 2666–2675.
Goldman, A. I. (1999). *Knowledge in a social world.* Oxford University Press.
Guiot, D., & Roux, D. (2010). A second-hand shoppers' motivation scale: Antecedents, consequences, and implications for retailers. *Journal of Retailing, 86*(4), 355–371.
Hair, J. F., Jr., Matthews, L. M., Matthews, R. L., & Sarstedt, M. (2017). PLS-SEM or CB-SEM: Updated guidelines on which method to use. *International Journal of Multivariate Data Analysis, 1*(2), 107–123.
Hemonnet-Goujot, A., Kessous, A., & Magnoni, F. (2022). The effect of sustainable product innovation on the consumer–luxury brand relationship: The role of past identity salience. *Journal of Business Research, 139*, 1513–1524.
Henseler, J., Ringle, C. M., & Sarstedt, M. (2015). A new criterion for assessing discriminant validity in variance-based structural equation modeling. *Journal of the Academy of Marketing Science, 43*, 115–135.
Holbrook, A. L., & Krosnick, J. A. (2010). Social desirability bias in voter turnout reports: Tests using the item count technique. *Public Opinion Quarterly, 74*(1), 37–67.
Holman, R. H. (1980). Clothing as communication: An empirical investigation. *Advances in Consumer Research, 7*(1).
Jaikumar, S., Singh, R., & Sari, A. (2018). "I show off, so I am well off": Subjective economic well-being and conspicuous consumption in an emerging economy. *Journal of Business Research, 86*, 386–393.
Jain, S. (2019). Factors affecting sustainable luxury purchase behavior: A conceptual framework. *Journal of International Consumer Marketing, 31*(2), 130–146.
Jain, S., & Rathi, R. (2023). Do consumer values and perceived readiness impact second-hand luxury consumption? A goal-framing theory approach. *Journal of Product and Brand Management, 32*(7), 973–987.
Jaiswal, D., & Kant, R. (2018). Green purchasing behaviour: A conceptual framework and empirical investigation of Indian consumers. *Journal of Retailing and Consumer Services, 4*, 60–69.
Jan, M. T. (2022). Factors influencing the purchase of circular economy products: A comparative analysis of Malysia and Turkey. *International Journal of Business and Society, 23*(2), 802–819.

Janssen, C., Vanhamme, J., Lindgreen, A., & Lefebvre, C. (2014). The Catch-22 of responsible luxury: Effects of luxury product characteristics on consumers' perception of fit with corporate social responsibility. *Journal of Business Ethics, 119*, 45–57.

Kapferer, J. N. (2012). Abundant rarity: The key to luxury growth. *Business Horizons, 55*(5), 453–462.

Kapferer, J. N., & Bastien, V. (2009). The specificity of luxury management: Turning marketing upside down. *Journal of Brand Management, 16*, 311–322.

Kapferer, J. N., & Michaut-Denizeau, A. (2014). Is luxury compatible with sustainability? Luxury consumers' viewpoint. *Journal of Brand Management, 21*, 1–22.

Kapferer, J. N., & Michaut-Denizeau, A. (2019). Are millennials really more sensitive to sustainable luxury? A cross-generational international comparison of sustainability consciousness when buying luxury. *Journal of Brand Management*, 1–13.

Kessous, A., & Valette-Florence, P. (2019). From Prada to Nada: Consumers and their luxury products: A contrast between second-hand and first-hand luxury products. *Journal of Business Research, 102*, 313–327.

Kessous, A., Valette-Florence, P., & De Barnier, V. (2017). Luxury watch possession and dispossession from father to son: A poisoned gift? *Journal of Business Research, 77*, 212–222.

Ki, C. W. C., Li, C., Chenn, A. S., Chong, S. M., & Cho, E. (2024). Wise consumer choices in online secondhand luxury (OSHL) shopping: An integrated model of motivations, attitudes and purchase intentions for OSHL as wise, conspicuous and sustainable consumption. *Journal of Retailing and Consumer Services, 76.*

Kopytoff, I. (1986). The cultural biography of things: Commoditization as process. The social life of things: Commodities in cultural perspective, 68, 70–73.

Louis, D., Lombart, C., & Durif, F. (2021). Apport de la sensibilité aux prix à l'étude du «gap» entre intentions et comportements responsables. *Revue management et avenir, (3)*, 103–127.

Mandaric, D., Hunjet, A., & Vukovic D. (2022). The impact of fashion brand sustainability on consumer purchasing decisions. *Journal of Risk Financial Management, 15*, 176.

Mick, D. G., Burroughs, J. E., Hetzel, P., & Brannen, M. Y. (2004). Pursuing the meaning of meaning in the commercial world: An international review of marketing and consumer research founded on semiotics. *Semiotica, 152*(1/4), 1–52.

Milgram, B. L. (2004). Refashioning commodities: Women and the sourcing of secondhand clothing in the Philippines. *Anthropologica*, 189–202.

Mishra, S., Jai, S., & Pandey, R. (2023). Conspicuous value and luxury purchase intention in sharing economy in emerging markets: The moderating role of past sustainable behavior. *Journal of Global Fashion Marketing, 14*(1), 93–107.

Modi, A., Paul, J., & Patel, J. (2016). Predicting green product consumption using theory of planned behavior and resasoned action. *Journal of Retailing and Consumer Services, 29*, 123–134.

Mohr, L. A., Webb, D. J., & Harris, K. E. (2001). Do consumers expect companies to be socially responsible? The impact of corporate social responsibility on buying behaviour. *Journal of Consumer Affairs, 35*(1), 45–72.

Moser, A. K. (2015). Thinking green, buying green? Drivers of pro-environmental purchasing behavior. *Journal of Consumer Marketing, 32*(3).

Nunnally, J. C. (1978). An overview of psychological measurement. In B. B. Wolman (Ed.), *Clinical diagnosis of mental disorders: A handbook* (pp. 97–146). Springer.

O'Cass, A., & McEwen, H. (2004). Exploring consumer status and conspicuous consumption. *Journal of Consumer Behaviour, 4*(1), 25–39.

O'Reilly, L., Rucker, M., Hughes, R., Gorang, M., & Hand, S. (1984). The relationship of psychological and situational variables to usage of a second-order marketing system. *Journal of the Academy of Marketing Science, 12*, 53–76.

Oppenheim, A. N. (1992). Questionnaire design. *Interviewing and Attitude Measurement, 24.*

Osburg, V. S., Davies, I., Yoganathan, V., & McLeay, F. (2021). Perspectives, opportunities and tensions in ethical and sustainable luxury: Introduction to the thematic symposium. *Journal of Business Ethics, 169*, 201–210.

Özçağlar-Toulouse, N. (2009). What meaning do responsible consumers give to their consumption? An approach by narratives. *Recherche et Applications en Marketing (English Edition), 24*(3), 3–20.

Preeti, T., Pareek, A., & Rishi, B. (2017). Social comparison of luxury fashion brands: Impact of ostentation and media habits. *Journal of Marketing Management, 16*(1).

Rawls, J. A. (1999). *Theory of Justice* (revised ed, p. 8). Harvard University Press.

Ringle, C., Da Silva, D., & Bido, D. (2015). Structural equation modeling with the SmartPLS. Bido, D., da Silva, D., & Ringle, C. (2014). Structural equation modeling with the Smartpls. *Brazilian Journal of Marketing, 13*(2).

Robeyns, I. (2008). Ideal theory in theory and practice. *Social Theory and Practice, 34*(3), 341–362.

Roux, D., & Guiot, D. (2008). Measuring second-hand shopping motives, antecedents and consequences. *Recherche et Applications en Marketing, 23*, 64–84.

Schultz, P. W., Gouveia, V. V., Cameron, L. D., Tankha, G., Schmuck, P., & Franěk, M. (2005). Values and their relationship to environmental concern and conservation behavior. *Journal of Cross-Cultural Psychology, 36*(4), 457–475.

Sempels, C. (2009). Vers une meilleure compréhension des écarts entre discours et comportements dans le cadre d'une consommation responsable: apports de la littérature. *8th International conference marketing-trends*, Venice.

Shukla, P. (2010). Status consumption in cross-national context: SocioPsychological, brand and situational antecedents. *International Marketing Review, 27*(1), 108–129.

Stolz, K. (2022). Why do (n't) we buy second-hand luxury products? *Sustainability, 14*(14), 1–24.

Stone, J., Home, S., & Hibbert, S. (1996). Car boot sales: A study of shopping motives in an alternative retail for- mat. *International Journal of Retail and Distribution Management, 24*(11), 4–15.

Taylor, S., & Todd, P. (1995). An integrated model of waste management behavior: A test of household recycling and composting intentions. *Environment and Behavior, 27*(5), 603–630.

Tu, J. C., Hsu, C. F., & Creativani, K. (2022). A study on the effects of consumers' perception and purchasing behavior for second-hand luxury goods by perceived value. *Sustainability, 14.*

Turunen, L. L. M., & Leipamaa-Leskinen, H. (2015). Pre-loved luxury: Identifying the meanings of second-hand luxury possessions. *Journal of Product and Brand Management, 24*(1), 57–65.

Turunen, L. L. M., & Pöyry, E. (2019). Shopping with the resale value in mind: A study on second-hand luxury consumers. *International Journal of Consumers Studies, 43*, 549–556.

Vigneron, F., & Johnson, L. W. (2004). Measuring perceptions of brand luxury. *Journal of Brand Management, 11*(6), 484–506.

Wattanasuwan, K. (2005). The self and symbolic consumption. *Journal of American Academy of Business, 6*(1), 179–184.

Weil, C. (1999). *Secondhand Chic: Finding fabulous fashion at consignment, vintage, and thrift stores*. Pocket Books.

Xu, Y., Chen, Y., Burman, R., & Zhao, H. (2014). Second-hand clothing consumption: A cross-cultural comparison between American and Chinese young consumers. *International Journal of Consumer Studies, 38*(6), 670–677.

Zeithaml, V. A. (1988). Consumer perceptions of price, quality, and value: A means-end model and synthesis of evidence. *Journal of Marketing, 52*(3), 2–22.

Zhang, L., & Cude, B. J. (2018). Chinese consumers' purchase intentions for luxury clothing: A comparison between luxury consumers and non-luxury consumers. *Journal of International Consumer Marketing, 30*, 336–349.

Agnès Walser-Luchesi is an Associate Professor of Marketing, PhD in Management Sciences—Marketing, accredited to supervise research at EM Strasbourg Business School, University of Strasbourg, France. She is a permanent member of the HuManiS (Humans and Management in Society) research center.

Her research focuses on price analysis and perceptual phenomena, the use of animals in advertising—drawing on anthropomorphism theory—as well as hospitality/gratitude mechanisms and customer listening practices within companies (such as corporate visits/museums, CEM, Affinity Diagrams, and Scenario Planning). She is the author of several teaching cases published with CCMP and is the head of the Master's degree in Marketing and Market Listening at EM Strasbourg. She teaches in various Master's programs and the Grande École program.

Previously, she held a position in the marketing department of an oil distribution company and also served as a project manager for the University of Strasbourg.

Anne-Catherine Furst a professional with over 20 years of experience within an international group in the field of medical diagnostics, I began my career as a specialized translator before advancing to roles in general administration, communication, and event management. In 2022, I enhanced my expertise with a Master's degree in Marketing and Market Research. Currently, I manage general administration, communication, and event organization for the EMEA region.

Landisoa Rabeson received her PhD in Economic and Social Sciences from the University of Fribourg, Switzerland. Currently, she is working as an Associate Professor of Marketing at EM Strasbourg Business School, University of Strasbourg, France. Her research focuses on consumer behavior, sustainability and service recovery which was published in Journal of Brand Management, Journal of Retailing and Consumer Services, Journal of Services Marketing, and Sustainability.

Chapter 11
Circular Economy in Practice: The Benefits of Collaboration for Securing Material Flow in a US Study

Daniel Borner and Barbara Eisenbart

Abstract Purpose—This paper aims to describe an amiable approach of addressing challenges to transitioning to a circular economy through collaboration and knowledge sharing between different partners.

Design/methodology/approach—The design science research methodology is used to identify the problem of the circular economy, suggest solutions, develop a usable artefact based on gathered user feedback interviews (n = 15) with US recyclers and industry experts.

Findings—There are two main approaches to navigating the circular economy, soft and hard methods. *Hard methods* include investing in **digitalization** and newer **technologies** to improve one's processes. However, these methods require an initial capital investment and high operation costs, which can deter many SMEs. *Soft methods* on the other hand, include methods such as **collaboration and sharing of knowledge**. These methods require no initial investments, but rather rely on the use of learning from partners to improve their processes and gather data to evaluate internal structures. These soft methods are explained in detail and an artefact has been developed to showcase these approaches.

Business and social implications—This article furthers ideas of the circular economy regarding collaboration to identify the right sourcing partner. This is of growing importance to meet regulatory requirements of increasing recycling quotas on specific materials like secondary plastic.

Originality/value—The developed artefacts of **material sheet** and **utility analysis tools** improve the transparency of reusable material flows. It has been proven for a global producer of reusable plastic packaging solutions operating in the US with state-specific recycling content requirements. It provides a decision-framework to secure sourcing of secondary plastic to acceptable costs, CO_2 reduction/sustainability and partner's fit.

D. Borner (✉) · B. Eisenbart
School of Business, University of Applied Sciences and Arts Northwestern Switzerland (FHNW), Windisch, Switzerland
e-mail: dannyborner@bluewin.ch; barbara.eisenbart@fhnw.ch

A. H. Verkuil et al. (eds.), *Core Values and Decision-Making for Sustainable Business*, Sustainable Business Development,
https://doi.org/10.1007/978-3-031-78361-6_11

Keywords Circular economy · Material flow · Recycling material streams · Sustainable sourcing · SME

Introduction

Businesses are increasingly confronted with the challenge of transitioning their business model to a more sustainable one. This challenge is accelerated by the rising prices of foreign primary materials and the need to adapt one's material flows to mitigate resource costs (Weber & Stuchtey, 2019). These types of challenges are common for enterprises and signify a paradigm shift towards the philosophy of **circular economy (CE)**. This is defined by Guldmann (2017) as "a systems model in which every part of a product is considered a valuable resource that should be carefully handled from cradle to cradle. This extends not only to the traditional supply chain, but also beyond that to consider raw material sourcing and disposal". However, there is still a misconception over what CE embodies as it is not an end goal, but rather a means to an end (Weber & Stuchtey, 2019). Therefore, CE must be seen as a subset of sustainability which has the potential to provide ecological and financial incentives for businesses willing to adopt sustainable business models. As regulations are increasingly being implemented requiring more recycling content, the subsequent price increase in primary plastic materials is shaping new long-term outlooks on circularity (Jonker et al., 2022). This shift in responsibility from simple raw material recovery to producer ownership is directly resulting in the increase in secondary plastic demand across impacted regions.

Figure 11.1 shows the changing responsibilities in circularity.

Theoretical Relevance

Underlying problems for CE are very intricate and difficult to surmise without a proper understanding of the theory behind CE. It is often not easy for businesses to

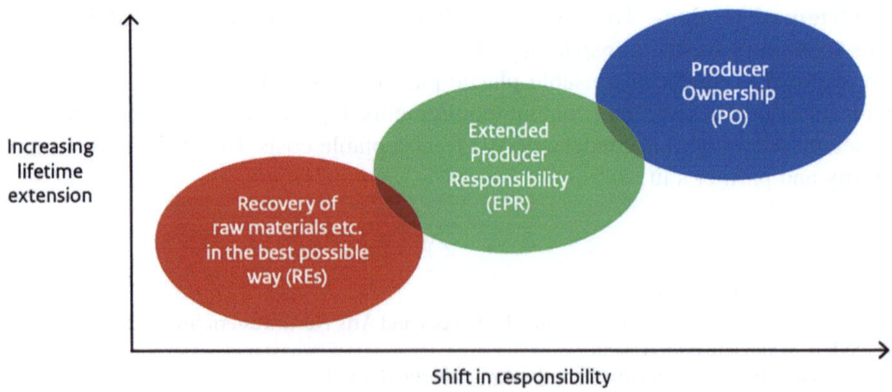

Fig. 11.1 Changing responsibilities in circularity; Source: Jonker et al. (2022)

understand the value of the CE and therefore many businesses do not shape their business models around this concept (Ellen MacArthur Foundation, 2013). The core principle of CE and sustainable strategies revolve around creating and retaining value. Through different types of actions, such as refurbishment, remanufacturing, reuse, and recycling, manufacturers and recyclers alike can benefit greatly (Reim et al., 2021). Adopting circular business models allows manufacturers and recyclers to manage their product lifecycles and thus empower their value chains with increased collaboration, knowledge, and transparency (Daaboul et al., 2016).

Business Relevance

Due to various economic and environmental factors, there is a growing need for businesses to adopt circularity into their business models (Lüdeke-Freund et al., 2019). Manufacturing companies, however, might hesitate as they are waiting for raw material prices to reach a lower price level. This is an unfortunate development as there is a strong indication that an economic adaptation of the CE could lead to net material cost savings (Guldmann, 2017).

Only 9% of plastic waste is recycled globally while 22% is mismanaged (OECD, 2022). In the US, most plastic waste is landfilled. According to the Organization for Economic Co-operation and Development, the United States of America is the leading country in per capita plastic waste generation. Companies need to meet regulatory requirements of increasing recycling quotas on specific materials like secondary plastic. However, recycled plastic is limited with high cost and limited availability.

Figure 11.2 shows global plastic waste and recycling.

This presents a challenge, but also an opportunity to address these recycling material streams on a very large scale. The recycling and reuse of raw materials and manufactured goods could provide manufacturing firms as well as waste management companies with the ability to reduce their production costs. This in turn would yield a competitive advantage due to the reduction of resource dependency on foreign primary and secondary materials (Weber & Stuchtey, 2019).

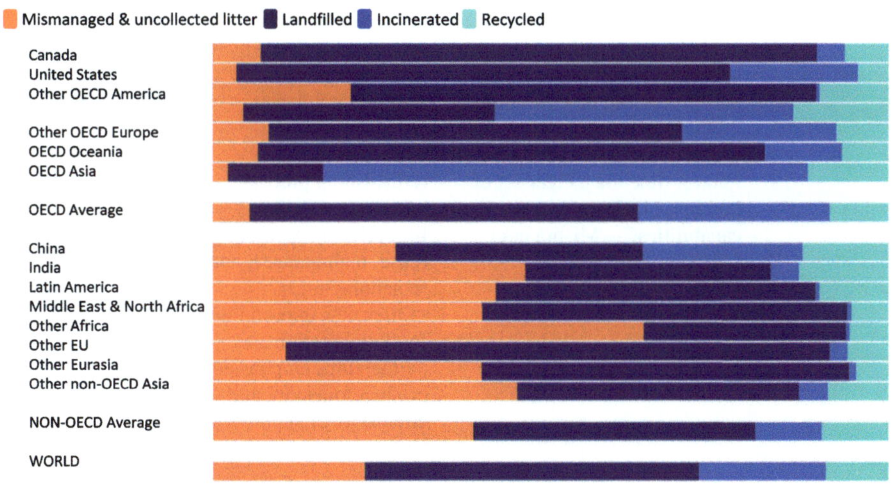

Fig. 11.2 Global plastic waste and recycling; Source: OECD (2022)

Key Concepts on Material Flows

Different articles have done a systematic review of the literature of circular economy and its related concepts, reflecting the ongoing importance of innovation and change within the circular economy ecosystem. In Table 11.1 we give an overview of key concepts from literature review articles.

Research Gap

In the current literature, there is a lack of case studies that analyze circular business model insights on the US recycling market due to various reasons such as lack of data, **collaboration**, and transparency (Heller et al., 2020). Studies as done by Reim et al. (2021) focus solely on large **Swedish manufacturing** companies and fail to address one of the biggest potential recycling markets worldwide. Admittedly stated, a broader case selection would provide scope for better cross-case analysis for future studies in the field (Reim et al., 2021). Additionally, Azizi et al. (2023) adds that **unstandardized methodologies** in this field leads to data limitations and narrow scope, reducing the significance of relevant and reliable information across regions.

The circular economy proposes strategies to retain and create value of resources (Garcia-Saravia Ortiz-de-Montellano & van der Meer, 2022). With new sources of value creation, closed-loop systems have the potential to offer a competitive advantage by providing cost advantages and creating new value (Ghisellini et al., 2016). By improving collaboration and reducing **data uncertainty** along the recycling

Table 11.1 Overview of key concepts from literature review articles; Source: Own contribution

Key concepts	Quotes from literature	Opportunities	Authors
Sustainable Sourcing	"The circular economy is a perspective with the explicit goal of **regenerating** natural, social, and economic capital in part by **cycling or cascading** products, parts and materials at their highest value for the longest time via a clear set of building blocks and capabilities."—Mishra et al. (2018)	• Optimize resource utilization and minimize waste generation which will lead to improved resource efficiency, reduced material consumption, and refurbishing processes. • Extend product lifecycles by promoting reuse, maintenance, and remanufacturing. • Offer new opportunities for collaboration and partnerships emphasizing product longevity and circularity of supply chains.	Ellen MacArthur Foundation (2013), Mishra et al. (2018), Reim et al. (2021)

(continued)

Table 11.1 (continued)

Key concepts	Quotes from literature	Opportunities	Authors
Recycling Value Chain	"Greater **coordination and cooperation** along the value chain is seen as essential to exploring and exploiting opportunities to increase plastic waste recycling."—Milios (2020)	• Recover and reuse valuable materials through the collection, refurbishment, and remanufacturing of products, leading to waste reduction. • Collaborate with stakeholders to share information, build networks, and increase knowledge to support each other with regulations and market demand. • Express requirements and facilitate better understanding of potential solutions for manufacturers and recyclers alike.	Milios (2020), Silva de Souza Lima Cano et al. (2022), Vlasopoulos et al. (2023)
Collaboration	"The recycling and dismissal activities could require and provide **information** on components, materials, and other resources through sharing and managing of data"—Terzi et al. (2010)	• Reduces the demand for primary resources and decreases waste generation. • Enables collaboration between, and informed decision making by, various stakeholders of a products lifecycle. • Provides an overview of the entire lifecycle of a product to identify the potentials for improving circularity.	Terzi et al. (2010), Daaboul et al. (2016), De Angelis et al. (2018), Takacs et al. (2022), Azizi et al. (2023)
Material Flow Analysis	"Material flow analysis is a systematic **assessment** of the state and changes of flows and stocks of **materials** within a system defined in space and time."—Brunner and Rechberger (2016)	• Leads to cost savings and resource efficiency through transparent assessment of processes and environmental impacts. • Addresses resource availability and potential scarcity by analyzing, planning, and allocating resources. • Offers insights into cost of waste, recycling/ treatment facilities, and product design in waste management.	Duygan and Meylan (2015), Brunner and Rechberger (2016), Heller et al. (2020)

(continued)

value chain, manufacturers and recyclers can lead to the realization of new and effective solutions (Azizi et al., 2023).

Alongside a study of the US market, data challenges in **plastic material flows** could be addressed, leading to improved data collection, coordination, and transparency (Heller et al., 2020). Therefore, the research gap which needs to be addressed is to improve the understanding of plastic material production using case studies to promote and further collaboration between plastic recycling market actors.

Data Collection & Analysis

The Design Science Research (DSR) methodology used in this project follows the four phases laid forth by Hevner and Chatterjee (2010) to create a framework that can analyze recycling material flows. Prior to delving into the DSR phases, the planned data collection and analysis for each of the phases is presented.

Research Design

The collection and analysis of data will be chronologically ordered alongside the four phases of the DSRM methodology. The primary methods of data collection include relevant literature, interviews with **recyclers and experts (n = 15),** and workshops with the Swiss and US sourcing teams. Throughout these interviews, data is collected to identify challenges and suggest strategies and solutions more accurately. All interviews are checked for accuracy by cross-referencing to preserve the integrity of the information collected. The goal is to identify key concepts from the plastic recycling industry and compare the results with findings from the literature to create an all-encompassing artefact.

Following the process of this model, questions per phase were identified and expected outcomes defined.

In Table 11.2 we illustrate a summary of design science research phases.

Phases for Artefact Development

The research followed the Design Science Research with explorative interviews (n = 15) of US recyclers and industry experts alike. The interviews can be broken down into the following: (n = 10) US recyclers, (n = 5) industry experts ranging from private sector, public sector, and non-profit organizations. Alongside these

Table 11.2 Summary of Design Science Research Phases; Source: Own contribution based on Hevner and Chatterjee (2010)

Phase	Research question per phase	Operational view	Expected outcomes
Awareness	What are the **challenges and difficulties** of securing plastic materials in the US market?	Identify what challenges recyclers and plastic manufacturers face in the US recycling market.	Identification of pain points including regulatory, consumer, and financial challenges. Understanding of financial issues concerning plastic recycling materials such as supply and demand, PCR versus PIR, and product specifications.
Suggestion	What **strategies and solutions** could help in securing limited supplies of secondary plastic material?	Identify solutions and strategies that could be implemented to secure more recycling material.	Development of first version of a suggested solution with phases for how to create the artefact effectively.
Development	How can the **artefact** be designed and developed to provide a practical overview of **relevant material flows** of potential partners?	Construct an artefact using collected data from recyclers with the ability to analyze flows and identify key partners along the value chain.	Development of prototype which incorporates all the variables needed to effectively determine the best procedure in securing plastic recycling material streams. Understanding of the key variables that need to be collected by the artefact to be of practical relevance to the case study.
Evaluation and Feedback	How effective is the developed artefact in addressing the identified problems and **improving material flow** transparency?	Test the constructed model alongside a US plastic manufacturer and receive user feedback to analyze the strengths and weaknesses of the artefact.	Evaluate the identified factors by gaining feedback from experts concerning the requirements from the awareness phase and artefact designed

interviews, workshops with stakeholders are conducted to refine the collected data and draw a consensus on the key factors identified. These workshops were organized into three sessions: first session with two experts from Swiss sourcing team, second session with two experts from US sourcing team, and third session with both Swiss and US project stakeholders.

The following Table 11.3 provides an overview of the data collected and analyzed between the two steps:

Table 11.3 Data collection and data analysis per research step; Source: Own contribution

	Awareness phase	Suggestion phase	Development phase	Feedback and evaluation
Data collection	Literature Review	Literature Review; Stakeholder Meeting	Workshop with the Swiss sourcing team and US sourcing team	Workshop with project stakeholders from both Swiss and US teams
	10 interviews with recyclers	5 interviews with industry experts	Send data sheet to potential recycling partners	
Data analysis	Identify pain points	Development of phases for artefact	Identifying key variables of soft and hard factors	Potential improvements of data sheet

Results

Qualitative Analysis

After thorough analysis of the literature, there are several concepts which have repeatedly been mentioned as relevant to the topic in discussion. These challenges include topics such as strategy, resources, knowledge, price, collaboration, legislation, design, and value. These challenges all relate to each other in several ways, but their conceptual definition remains unique. These criteria occur often in tandem and thus share similar pains and hurdles.

The ten interviews conducted with potential recyclers of interest for the US plastic manufacturer provided a basis for understanding the current pain points within the US case study. These ten recyclers are all located in the surrounding states of Indiana to represent the interests and knowledge reflected by the US plastic manufacturer. With the conducting of the interviews, the documentation was developed using the Atlas.ti software, and subsequently key concepts were identified. This input was refined by creating several codes that structured the concepts identified by the Atlas.ti tool into more coherent and fluent concepts. To reduce redundancy in the data, codes were developed out of the found concepts which resulted in 24 codes that bundled together very similar concepts. Figure 11.3 depicts the frequency of the codes distinguishes a clear picture of most addressed concepts. It is important to note that several concepts identified by the software tool such as "Question", "Speaker", and "Names" were omitted to avoid redundancy and data inaccuracy.

The five interviews conducted with industry experts within the plastic recycling industry were refined into specific codes representing the requirements of the artefact and its characteristics. These five experts cover a variety of industries from the public sector, private sector, and non-profit organizations (NGOs). As with the recycler interviews, Atlas.ti software was used for deeper analysis of the interviews. The findings in Fig. 11.4 represent a mostly balanced priority of characteristics with availability and efficiency of resources showing the largest discrepancy. As identified by various researchers, concepts such as scope (Reim et al., 2021), resources

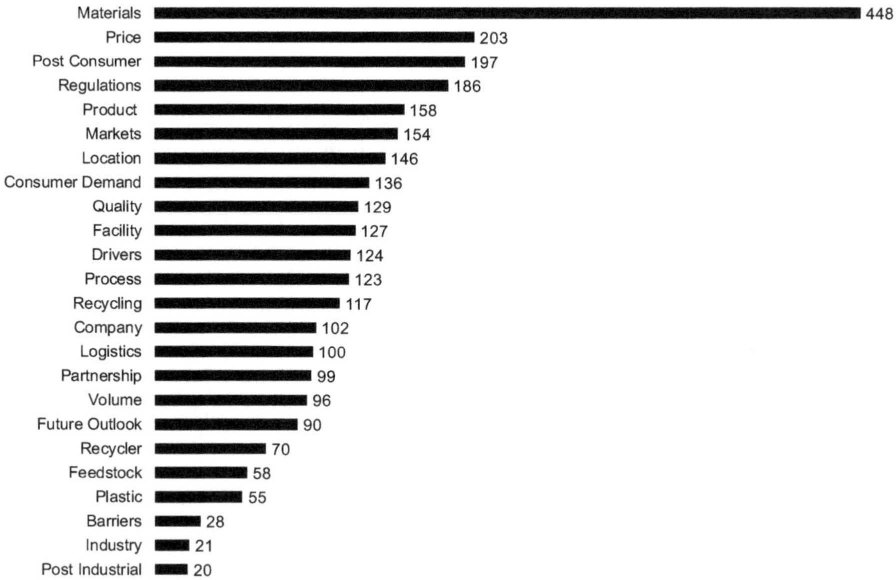

Fig. 11.3 Frequency of code applied-recycler interviews; Source: Own contribution

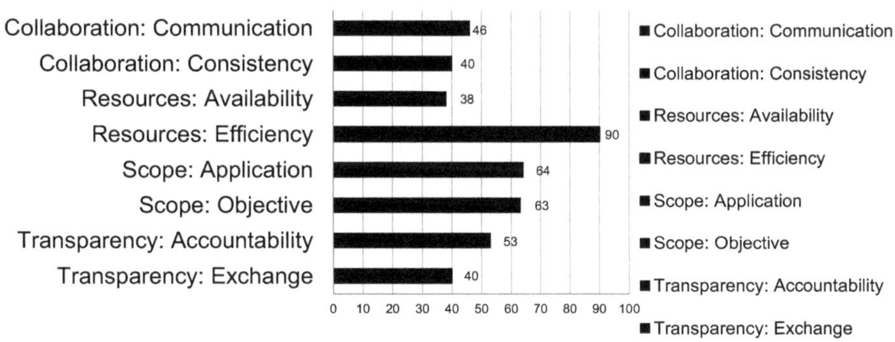

Fig. 11.4 Frequency of characteristics from expert interviews; Source: Own contribution

(Milios, 2020), transparency, and collaboration (Heller et al., 2020) play a significant role when breaking down the frequency of literature-based solutions. Combined with the five interviews conducted with industry experts, the frequency of these concepts can be interpreted to show a desire for a solution that can address resource efficiency and be specific to the scope in question. However, a balanced approach for creating an artefact that can address all characteristics seems to be most desirable. For specificity, these concepts are also further broken down into the two most frequent characteristics identified alongside the Atlas.ti software, leading to a concise examination of primary and secondary data.

Requirements for Artefact

The findings from both the primary and secondary data emphasize the importance of **materials/resources, pricing, regulations, and product quality** as challenges. Although there are similarities in thought, differences are also very evident. Results from these interviews reveal a difference in approach to priorities, namely concerning product design, value chains, product specifications, and location of facility/sourcing. If looking for a solution that reflects both values from secondary and primary data, requirements such as **scope, resources, collaboration, and transparency** can be broken down into more specific characteristics to identify key implications of the artefact.

The **scope** requirement can be broken down into the characteristics of application and objective. The application is specific to the design of the product, making it important to target applications within the scope of the artefact. It is equally significant to come up with an objective that the artefact must have since this will provide a structured approach to achieving circularity and thus securing material flows.

The **resource** requirement specifically mentions availability and efficiency as primary characteristics. These characteristics are crucial in not only identifying in a first step the availability of a more sustainable resource, but also the economic efficiency of a new resource. The availability of a resource in this context directly influences whether a more sustainable resource, such as post-consumer plastic, can be acquired in a big enough supply to maintain the current processing standards and targets. The efficiency of a resource impacts the quality and subsequent profitability of a product which was produced using the resource.

The **collaboration** requirement can be further analyzed by breaking it down into consistency and communication. These two characteristics of collaboration provide a balanced relationship between partners in that they address the consistency of delivery on materials and promote open communication. This leads to traceability along the production lines. Consistency is also very important within a partnership as it requires the supplier to deliver volumes of material regularly and the producer needs to take these resources off suppliers' hands, or the resources could be wasted. This indirectly requires communication to be effective to avoid any miscommunication concerning resource delivery, quality, and quantities needed.

The **transparency** requirement signifies two characteristics essential namely openness and accountability. Openness requires partners to be transparent, but also to be willing to change their business models and shift towards secondary plastic materials. For most recyclers and manufacturers, this requires a willingness for long-term investment and a shift in priority from linear to more circular thinking. This leads to accountability, as any tools or knowledge sharing which could allow for more traceability, would make market actors more accountable throughout their business processes.

Key Variables Identified

In a series of collaborative workshops, the development of a comprehensive artefact for data collection and utility analysis was undertaken with teams from Switzerland and the US. Participants, bringing extensive sales experience and insights into specific product qualities and quantities, worked on refining the material sheets and finalizing the weighted factors for analysis. Essential factors for data collection were analyzed, variables adjusted, and criteria for utility analysis weighted and rated. The final output included a thoroughly vetted material sheet and a robust utility analysis framework, ready for detailed examination of the results.

The **material sheets** and **utility analysis** can act as a basis for collecting data on potential partners. However, it is recommended to only collect basic information and not be overly complex. This leads to a breakdown into two requirements: soft and hard factors. Soft factors encompass collaboration and knowledge sharing, relying on partner learning to improve processes and gather data without needing initial investments. Hard factors involve investing in digitalization and new technologies to enhance processes, though they require significant initial capital and high operational costs.

Table 11.4 illustrates the final categories for soft and hard factors.

The **utility analysis** introduces a structured process to evaluate recycler inputs by factoring in various criteria. Each criterion is scored on a scale from one to five, with one indicating very poor performance and five indicating excellent performance. These scores are then aggregated and weighted according to their importance to US plastic manufacturers. An in-depth explanation is provided for each

Table 11.4 Final categories for soft and hard factors. Source: Own contribution

	Soft factors	Hard factors
Description	*Require exchange and disclosure* • Mutual understanding • Data sharing • Collaboration • Knowledge sharing	*Requires process improvements* • Initial capital investment • Operations investment • Invest in digitalization • Implementation of new technologies
Criteria	*Business Strategy* • Future plans (planned investments) • Partnerships (existing, planned) *Product Sourcing* • Sources of material • Geographic sources • Transportation • Technologies	*Processing and Applications* • Plastic material volume processed per month and year (in tons) • Use of production quality *Quantity and Quality* • Certifications for plastic material • Certifications for corporate social responsibility • Available (or needed) quantities of different plastic qualities per month (in lbs) • Specifications: Base quality, Melt flow rate, Impact modification, Stiffness modification, and colors

Categories	Criteria	Weighting (max. 100%)	Rating (1 low, 5 high)	Points
Material flows	Availability of secondary plastic materials (in volume)	90%	3	2.7

Fig. 11.5 Example of utility analysis criteria; Source: Own contribution

score to ensure transparency and traceability. These criteria are interchangeable and depending on user requirements, can be adjusted given the task needed to be fulfilled.

Figure 11.5 shows an example of utility analysis criteria.

Artefact Analysis

The artefact presents a variety of strengths and weaknesses for the US plastic manufacturer. Creating and identifying data points, such as available quantities and qualities of specific recyclers, provides the manufacturer with the ability to analyze and find recyclers that can fulfill the manufacturing requirements required for the short- and long-term. Also, understanding the priorities of recyclers provides a clear picture of future investments and capacities of volume when evaluating the possibilities of collaboration. The developed analysis tool provides the ability to rank potential collaborators and dynamically adjust weighted factors according to short- and long-term needs. However, it is crucial that the artefact remain dynamic to account for changes in needs, hence increasing the consistent effectiveness of the artefact.

The development and maintenance of the artefact can be strengthened through few changes. Prior to interviewing recyclers, providing a material sheet which is analyzed and developed in unison with recyclers would provide the artefact with a deeper understanding of available quantities and qualities. Expanding the search for appropriate recyclers can be done by conducting a pre-screening process. This would identify recyclers which score high on the highest weighted factors, such as use of injection molding, and result in a list of only the best possible recyclers for collaboration, prior to a more thorough analysis using the developed artefact. Further factors, such as consistency of delivery and willingness to adopt circular strategies, need to be included in the analysis of recyclers. This enables a more accurate alignment of economic and cultural tendencies of potential collaboration with US plastic manufacturers. The identified variables can become more detailed and thorough in their analysis with more input from third parties to not only reflect the needs of US plastic manufacturers. It would increase the scope of analysis tools to a wider range of plastic manufacturers across the US.

Discussion, Conclusion and Further Research

Findings

Several challenges were identified such as a lack of knowledge for newer solutions, lack of infrastructure for capturing waste streams, and lack of government incentives for more circular approaches. Reim et al. (2021) and De Angelis et al. (2018) conclude that the high cost of secondary materials and resource shortage are financial challenges that recyclers and plastic manufacturers both are currently faced with. Also, the ability to recapture value is not as simple as theorized, as certain plastic products are difficult to recapture due to their long product lifecycles. Consumer demand and regulations are forcing recyclers and plastic manufacturers alike to remain flexible to continuously offer the best product available for their end-consumers. This leads to a primary focus of recyclers and plastic manufacturers on profitability and cost savings regarding plastic materials. At the core of addressing these barriers, greater collaboration and coordination along the value chain will decrease operational losses and reduce external inputs, allowing for resource efficiency. This would allow for companies along the value chain to express their requirements explicitly and openly facilitating better solutions (Milios, 2020). Figure 11.6 describes the plastic waste recycling value chain.

Analysis of **material flows** can help identify the best potential partners for plastic manufacturers. According to information gathered throughout the workshops with both Swiss and US sourcing teams, the investment into newer technologies was also key. Such investments lead to an improved quantity and quality of products across the plastic recycling value chain. As Brunner and Rechberger (2016) and Daaboul et al. (2016) suggest, managing resource data and coordinating product lifecycles can enhance the necessary collaboration and partnership. This is because the sharing of information and optimization of processes can lead to more efficient practices, leading to less waste of manufacturing processes (Terzi et al., 2010). By sharing the burden of rising costs of secondary plastic materials market actors can also reduce the financial risk of transitioning to a CE.

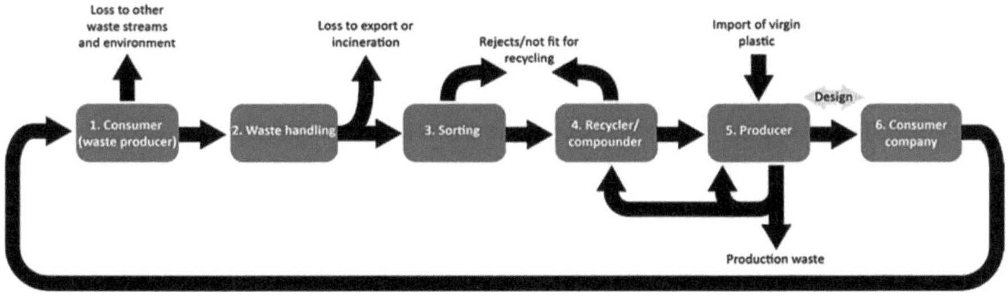

Fig. 11.6 Plastic waste recycling value chain; Source: Milios (2020)

Managerial Implications

The **material sheet** and **utility analysis tools** are the basis for identifying challenges and improving transparency of material flows. Due to the lack of available data of US recycling material flow streams, the data had to be cross-checked, and redundancies evaluated. It is essential to differentiate between flexibility of available qualities and consistency of delivery. This indicates a desire of manufacturers not only to understand what recyclers can provide, but how consistently they can provide certain materials. Ensuring a seamless collaboration and manufacturing process for the plastic manufacturer. According to Reim et al. (2021), investment into digitalization and maximizing value creation are the primary means for pursuing circular business models. However, contrary to this opinion, US plastic manufacturers do not have the infrastructure and advancements yet to support this. Therefore, they lean more towards collaboration and transparency as key factors for adopting circular business models.

The DSR methodology resulted in the creation of the artefact which consists of a material sheet and a utility analysis. Combined, they identify key variables and factors for the evaluation of **best suited sourcing partner** in the scope of the US plastic recycling market. Alongside the material sheet and utility analysis, plastic packaging manufacturers can gather data and analyze recyclers based on their short- and long-term needs. This comprehensive, data-driven assessment has proven to be effective in bridging the gap between players along the same value chain. The proper use of these tools allows for US plastic manufacturers to secure recycling flows through collaborative engagement such as aligning yearly production quantities, pre-ordering large quantities of material, and sharing short- and long-term investment plans. This will inevitably lead recyclers to reduce the burden of rising costs and an increasing shortage of secondary plastic materials. This, in turn, increases the circulation of resources and allows for the manufacturers and recyclers to pivot away from linear resource flows and adapt their business models to circular resource flows (Bocken et al., 2016).

Limitations

Although the understanding of circular resource flows is presented in the context of the US recycling market, the results can be generalized on a broader scale. As the concept of CE expands and consumer demand for sustainable products increases, other industries such as pharmaceuticals and retail sectors are also impacted. These markets also use recycling materials and through consumer demand and regulations will have to meet certain standards of sustainability (Silva de Souza Lima Cano et al., 2022). Their value chains can also benefit from tracking their own material flows through collaboration with the right partners and therefore, secure and increase the value of their own products.

As the developed artefact was created with a specific scope in mind, however, the artefact might pose limitations in its adaptation by different market sectors. Also, as this study analyzes the US recycling market, which is considered a developed country, plastic manufacturers and recyclers from less developed countries may not see the full value of the developed artefact. To adequately evaluate this concern, further case studies with a wider or different scope of reference are needed.

Impact and Transferability to Other Areas

Using the gathered inputs and conceptualizing their practical impact, not only for the scope in which it was investigated, provides examples of transferability to other areas. Across different industries, a similar principle of collecting performance data and conducting a utility analysis can provide any organization with great benefits. It can help analyze competitors and identify potential partners that one could work with. In lesser developed countries, promoting collaboration and coordination of business practices can help build infrastructure which will eventually lead to an increase in manufacturing capacity and plastic waste solutions. It is important to narrow a specific scope of reference when transferring the results of this case study, as it is core to be as specific as possible when identifying relevant data variables and analysis factors. However, the limited scope of reference for this case study can limit the possibilities of implementation across different markets and countries. As this study was solely focused on addressing the challenges and needs of the US recycling market, further studies in countries and markets with different waste and recycling infrastructure might come to different conclusions. Infrastructure will directly impact the limitations of manufacturing capacities due to a lack of advanced technology and the availability of non-bias, accurate data for interpretation.

Future Research

Finally, to fully understand the broader landscape of recycling markets, further studies must be conducted. This would generate more data and analyze the results identified for the US recycling market and expand upon the concept of CE. Further analysis should also be conducted into the impacts of product design through collaboration. The efficient design of products has already been proven to be an effective value proposition for circularity. Further research into the benefits of designing products in collaboration with actors across the value chain could pose even greater benefits for capturing and increasing the value of resources. Also, specifically for the US market, a more in-depth study of the current regulatory landscape of the US and impact of recycling content regulations is needed. Although these **regulations** are still only implemented in a minority of states, the increase in consumer demand will predictably lead to more regulations across the US.

References

Azizi, D. D. S., Hanafiah, M. M., & Woon, K. S. (2023). Material flow analysis in WEEE management for circular economy: A content review on applications, limitations, and future outlook. *Sustainability, 2023*(15), 3505. https://doi.org/10.3390/su15043505

Bocken, N. M. P., de Pauw, I., Bakker, C., & van der Grinten, B. (2016). Product design and business model strategies for a circular economy. *Journal of Industrial and Production Engineering, 33*(5), 308–320. https://doi.org/10.1080/21681015.2016.1172124

Brunner, P. H., & Rechberger, H. (2016). *Handbook of material flow analysis: For environmental, resource, and waste engineers* (2nd ed.). CRC Press. https://doi.org/10.1201/9781315313450

Daaboul, J., Le Duigou, J., Penciuc, D., & Eynard, B. (2016). An integrated closed-loop product lifecycle management approach for reverse logistics design. *Production Planning and Control, 27*(13), 1062–1077. https://doi.org/10.1080/09537287.2016.1177234

De Angelis, R., Howard, M., & Miemczyk, J. (2018). Supply chain management and the circular economy: Towards the circular supply chain. *Production Planning and Control, 29*(6), 425–437. https://doi.org/10.1080/09537287.2018.1449244

Duygan, M., & Meylan, G. (2015). Strategic management of WEEE in Switzerland - Combining material flow analysis with structural analysis. *Resources, Conservation and Recycling, 103*, 98–109. https://doi.org/10.1016/j.resconrec.2015.06.005

Ellen MacArthur Foundation. (2013). *Towards the circular economy, Vol. 1*. https://emf.thirdlight.com/link/x8ay372a3r11-k6775n/@/preview/1?o

Garcia-Saravia Ortiz-de-Montellano, C., & van der Meer, Y. (2022). A theoretical framework for circular processes and circular impacts through a comprehensive review of indicators. *Global Journal of Flexible Systems Management, 23*(2), 291–314. https://doi.org/10.1007/s40171-022-00300-5

Ghisellini, P., Cialani, C., & Ulgiati, S. (2016). A review on circular economy: The expected transition to a balanced interplay of environmental and economic systems. *Journal of Cleaner Production, 114*, 11–32. https://doi.org/10.1016/j.jclepro.2015.09.007

Guldmann, E. (2017). Best practice examples of Circular Business Models. *Danish Environmental Protection Agency*. https://doi.org/10.13140/RG.2.2.33980.95360

Heller, M. C., Mazor, M. H., & Keoleian, G. A. (2020). Plastics in the US: Toward a material flow characterization of production, markets and end of life. *Environmental Research Letters, 15*(9). https://doi.org/10.1088/1748-9326/ab9e1e

Hevner, A., & Chatterjee, S. (2010). *Design research in information systems* (Vol. 22). Springer US. https://doi.org/10.1007/978-1-4419-5653-8

Jonker, J., Faber, N., & Whitepaper, T. H. (2022). *Circular business models. A study to classify existing and emerging forms of value retention and creation*. https://research.hanze.nl/ws/portalfiles/portal/42884440/Whitepaper_Circular_Business_Models_ebook.pdf

Lüdeke-Freund, F., Gold, S., & Bocken, N. M. P. (2019). A review and typology of circular economy business model patterns. *Journal of Industrial Ecology, 23*(1), 36–61. Blackwell Publishing. https://doi.org/10.1111/jiec.12763.

Milios, L. (2020). *Policy framework for material resource efficiency: Pathway towards a circular economy*. https://portal.research.lu.se/en/publications/policy-framework-for-material-resource-efficiency-pathway-towards

Mishra, J. L., Hopkinson, P. G., & Tidridge, G. (2018). Value creation from circular economy-led closed loop supply chains: A case study of fast-moving consumer goods. *Production Planning and Control, 29*(6), 509–521. https://doi.org/10.1080/09537287.2018.1449245

OECD. (2022). *Global plastics outlook*. OECD Publishing. https://doi.org/10.1787/de747aef-en

Reim, W., Sjödin, D., & Parida, V. (2021). Circular business model implementation: A capability development case study from the manufacturing industry. *Business Strategy and the Environment, 30*(6), 2745–2757. https://doi.org/10.1002/bse.2891

Silva de Souza Lima Cano, N., Iacovidou, E., & Rutkowski, E. W. (2022). Typology of municipal solid waste recycling value chains: A global perspective. *Journal of Cleaner Production, 336*. https://doi.org/10.1016/j.jclepro.2022.130386

Takacs, F., Brunner, D., & Frankenberger, K. (2022). Barriers to a circular economy in small- and medium-sized enterprises and their integration in a sustainable strategic management framework. *Journal of Cleaner Production, 362*. https://doi.org/10.1016/j.jclepro.2022.132227

Terzi, S., Bouras, A., Dutta, D., Garetti, M., & Kiritsis, D. (2010). Product lifecycle management - From its history to its new role. *International Journal of Product Lifecycle Management, 4*(4), 360–389. https://doi.org/10.1504/IJPLM.2010.036489

Vlasopoulos, A., Malinauskaite, J., Żabnieńska-Góra, A., & Jouhara, H. (2023). Life cycle assessment of plastic waste and energy recovery. *Energy, 277*. https://doi.org/10.1016/j.energy.2023.127576

Weber, T., & Stuchtey, M. (2019). *Deutschland auf dem Weg zur circular economy*. https://www.acatech.de/publikation/deutschland-auf-dem-weg-zur-circular-economy/

Daniel Patrick Borner holds a Bachelor of Arts in International/Global Studies from The College of New Jersey, USA, and a Master of Science in Business Information Systems from the University of Applied Sciences and Arts Northwestern Switzerland. He is currently working within the Planning and Logistics Chapter at Roche Basel, Switzerland, applying and expanding his knowledge in operations and strategic management. Previously, he was Planning Manager at the Swiss Armed Forces and Web Developer at Dorner Health IT Solutions.

Barbara Eisenbart, Prof. Dr., is an economist with expertise in entrepreneurial thinking, new venturing and transformation. She holds a doctorate in business innovation at University of St. Gallen, Switzerland. She studied at Harvard University and has a Master of Economics from Sophia University, Tokyo as well as an MBA from University of Göttingen, Germany. She is Professor at the Institute of Management at School of Business, University of Applied Sciences and Arts Northwestern Switzerland. Barbara Eisenbart has more than 20 years of experience in working in companies. She was responsible for global transformation projects at Schindler and business development at the agribusiness Syngenta for renewables with market in Brazil. Also, she worked in strategic consulting at McKinsey USA and Germany.

Chapter 12
Final Thoughts: Entrepreneurship Challenges in a Demanding Political and Economic Ecosystem

Uta Milow

We saw in the chapters of this book that small and medium-sized enterprises (SMEs) face many challenges in today's business landscape, in particular competitive pressures, societal expectations, and economic and political uncertainties.

Competition and Cost Pressures

In an increasingly globalized and digital economy, SMEs are not only competing with local businesses but also with large multinational corporations. These larger companies benefit from economies of scale and advanced technology, making it difficult for SMEs to offer the same prices or range of services. Rising input costs due to higher inflation rates from raw materials to labor, reduce profit margins. The post-pandemic economy has seen supply chain disruptions that continue to affect SMEs, making it harder for them to maintain cost efficiency while delivering products or services.

Stakeholder Pressure for Sustainability

On the other hand, the demand for sustainable business practices is growing. Stakeholders, like customers, investors, and regulators, increasingly expect SMEs to adopt environmentally friendly and socially responsible business operations. This can be especially challenging for smaller companies, as implementing green

U. Milow (✉)
University of Applied Sciences and Arts Northwestern Switzerland, Windisch, Switzerland
e-mail: Uta.milow@fhnw.ch

A. H. Verkuil et al. (eds.), *Core Values and Decision-Making for Sustainable Business*, Sustainable Business Development,
https://doi.org/10.1007/978-3-031-78361-6_12

technologies, using sustainable materials, and ensuring fair supply chains often require significant investment. While these measures can lead to long-term benefits and better brand reputation, the initial costs and complexity are barriers for many SMEs already facing financial strain.

Nevertheless, SMEs must address shifting consumer needs. Customers today prioritize companies that align with their values, demanding more than just products—they seek brands that contribute to society. This has led SMEs to offering products that take health-consciousness, eco-awareness, and ethical aspects into account. While this provides an opportunity for SMEs to differentiate themselves, meeting these demands requires innovation and resources.

Uncertainties in the Entrepreneurial Ecosystem

The ongoing war in Ukraine has caused significant geopolitical instability, disrupting global trade and creating uncertainty in energy markets, especially in Europe. The war has also caused shifts in political alliances, trade sanctions, and new regulations, which add further uncertainty to SMEs operating in affected regions or sectors. It is important to realize that geopolitical risks have increased in recent years and have been weighing on the global economy for quite some time. The antagonism between the West and Russia and China has increasingly divided the international community of states, significantly limiting or even preventing cooperation. Concern over the increasing rivalry between the US and China is also leading to uncertainty. Should domestic demand for consumer goods in China continue to slump, further measures to boost exports can be expected. According to analysts, this will further fuel the rivalry regardless of the outcome of the US presidential elections (Kennedy, 2024). On the other hand, the USA is currently the growth engine of the global economy. For the current year, the US Department of Commerce's Bureau of Economic Analysis expects the American economy to grow by 2.7%—1.2 percentage points more than forecast in the fall. And next year, the economy is still expected to grow by 1.9% (*Gross Domestic Product U.S.*, 2024).

Economic uncertainty remains one of the most pressing challenges for SMEs. Inflationary pressures with the central banks raising interest rates (*IMF World Economic Outlook (April* 2024*)*, 2024) increase borrowing costs for businesses. At the same time, consumption declines as consumers face their own financial pressures. For SMEs, already more vulnerable to economic shocks than larger firms, these uncertainties can be a make-or-break factor.

In sum, today's SMEs face a set of challenges that require resilience, adaptability, and innovation. The pressures of competition, stakeholder expectations, and economic volatility around sustainability need strategic responses. SMEs that can adapt quickly, implement digital tools, and align their business models with evolving societal and economic trends will be better positioned to survive. However, without the right support, such as access to finance and favorable policy environments, many SMEs may struggle in this volatile landscape.

There is more to analyze and explore – following volumes of this series will continue the discussion.

References

Gross Domestic Product U.S. (2024). U.S. Bureau of Economic Analysis (BEA). https://www.bea.gov/data/gdp/gross-domestic-product

IMF World Economic Outlook (April 2024*)*. (2024). https://www.imf.org/external/datamapper/PCPIPCH@WEO/OEMDC/ADVEC/WEOWORLD

Kennedy, S. (2024). *U.S.-China relations in* 2024*: Managing competition without conflict*. https://www.csis.org/analysis/us-china-relations-2024-managing-competition-without-conflict

Uta Milow, Prof. Dr., is a lecturer and researcher of economics at FHNW with a research focus on sustainable and international entrepreneurship. She attended the University of Mannheim, Germany, and the University of Oregon, USA. After finishing a master's degree in economics at the University of Mannheim, she continued her studies with a PhD-program in Economics at Eberhard-Karls-University Tübingen, Germany. She participated in an interdisciplinary postgraduate research program on European integration (political science, law, and economics). Since 2000 she has been a lecturer at the School of Business FHNW. She is head of the Master of Business Administration (MBA) programs at the School of Business, FHNW, and is responsible for all executive education programs of the Institute of Management, FHNW.

MIX
Papier aus verantwortungsvollen Quellen
Paper from responsible sources
FSC® C105338
www.fsc.org

If you have any concerns about our products,
you can contact us on
ProductSafety@springernature.com

In case Publisher is established outside the EU,
the EU authorized representative is:
Springer Nature Customer Service Center GmbH
Europaplatz 3, 69115 Heidelberg, Germany

Printed by Libri Plureos GmbH
in Hamburg, Germany